T0189883

Love's Rite

Same-Sex Marriage in India and the West

Ruth Vanita

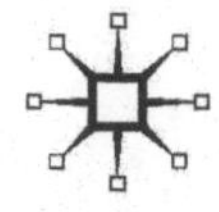

LOVE'S RITE

Softcover reprint of the hardcover 1st edition 2005 978-1-4039-7038-1

First published in 2005 by
PALGRAVE MACMILLAN™
175 Fifth Avenue, New York, N.Y. 10010 and
Houndmills, Basingstoke, Hampshire, England RG21 6XS
Companies and representatives throughout the world.

PALGRAVE MACMILLAN is the global academic imprint of the Palgrave Macmillan division of St. Martin's Press, LLC and of Palgrave Macmillan Ltd. Macmillan® is a registered trademark in the United States, United Kingdom and other countries. Palgrave is a registered trademark in the European Union and other countries.

ISBN 978-1-349-53208-7 ISBN 978-1-4039-8160-8 (eBook)
DOI 10.1057/9781403981608

Library of Congress Cataloging-in-Publication Data

Vanita, Ruth.
Love's rite : same-sex marriage in India and the west / by Ruth Vanita.
p. cm.
Includes bibliographical references and index.

1. Same-sex marriage. 2. Same-sex marriage—India. 3. Same-sex marriage—United States. I. Title.

HQ1033.V36 2005
306.84'8—dc22 2005047571

A catalogue record for this book is available from the British Library.

Design by Newgen Imaging Systems (P) Ltd., Chennai, India.

First edition: November 2005

10 9 8 7 6 5 4 3 2 1

Transferred to Digital Printing in 2009

For Mona

Kama, God of Love, shooting an arrow at two women. Sandstone sculpture, Orissa, eastern India, ca. eleventh century. Seattle Art Museum, Accession no. 74.17. Photo by Paul Macapia.

Ishq par zor nahin, hai yeh voh atish Ghalib
Ki lagaye na lagen aur bujhaye na baney

(Love cannot be forced—it is a flame, says Ghalib,
Which cannot be ignited at will or quenched at will)

The perfect ceremony of love's rite

—*William Shakespeare*, Sonnet 23

Contents

List of Illustrations

Acknowledgments

A book of this kind draws on so many conversations and acts of kindness over the years that it is impossible to remember and list them all. Some are acknowledged in footnotes; if I have inadvertently omitted anyone, I apologize.

I gratefully acknowledge the support of the National Endowment for the Humanities, the American Council of Learned Societies, and the Social Science Research Council, who awarded me an ACLS/SSRC/NEH International and Area Studies Fellowship for the year 2003–2004, which enabled me to complete this book. My views, findings, conclusions, and recommendations do not necessarily represent those of the National Endowment for the Humanities.

I am thankful to the University of Montana for granting me unpaid leave for 2003–2004, and allowing me to postpone my sabbatical leave to 2004–2005, thus enabling me to accept the fellowship. I am grateful to my colleagues in the Liberal Studies Program, Paul Dietrich, Alan Sponberg, and Stewart Justman, for their ever-cheerful help and warm collegiality, to Provost Lois Muir for her support, and also to friends in other departments, especially Bruce Bigley, Michael Mayer, G.G. Weix, and Shiv Ganesh. Terry Castle, Lillian Faderman, and Eileen Barrett, thank you for generosity of many kinds over many years.

I am very grateful to Swami Bodhananda Saraswati for granting me an interview. I thank all the individuals and couples, named and unnamed, who shared their stories with me, and those who agreed to let me reproduce their photographs. I am grateful to *The Week* for their photograph of Santosh and Manju, and especially to Litta Jacob for helping me obtain it; and to *Savvy* for their photographs of Leela Namdeo and Urmila Srivastava, and of Sheela and Sree Nandu. Thanks to National Film Archive of India for the film still from *Dosti*.

An outline of my argument in this book appeared as an essay, " 'Wedding of Two Souls': Same-Sex Marriage and Hindu Traditions," in *Journal of Feminist Studies in Religion* 20: 2 (Fall 2004). Some materials from chapter 8 appeared in expanded form in an essay, "Married Among their Companions: The Representation of Female Homoerotic Relations in Nineteenth-Century Urdu *Rekhti* Poetry," in *Journal of Women's History* 16:1 (Spring 2004), 12–53. I am grateful to readers of these journals, who commented on these essays.

I presented materials from various chapters, while this book was a work in progress, at many venues, including Hindu College, Delhi University; Women's Studies and South Asia Studies Programs, University of Michigan; Massachusetts Institute of Technology; South Asia Program, Cornell University; Center for Lesbian and Gay Studies at the City University of New York; Center for the Study of Gender and Sexuality at NYU; Women's Studies Program, Seattle University; the Pacific Asian and North American Asian Women in Theology and Ministry annual conference, 2004; Bard College "Homosexuality and World Religions" conference, 2003; Swarthmore College; Purchase College; Kalamazoo College; and Hobart and William Smith College. I am grateful to the organizers of these talks, and to the audiences whose questions and comments enriched my work.

Many thanks to Madhav Deshpande, Professor at the University of Michigan, for teaching me Sanskrit, and for responding to my questions and helping me translate certain phrases. I am grateful to Darshan Singh Kang (retired professor, University of Montana) for his extended assistance while I was learning to read and translate Urdu poetry, and to Jeety Kang for loving hospitality during our reading sessions. Thanks to Anannya Dasgupta for translating from Bengali; to Geeta Patel for insightful conversations; to Sandip Roy, editor of *Trikone*, who always responds promptly to my queries; to members of the Khush listserv and the LGBT Hindu listserv for directing me to several sources; to Bridget Whearty for research assistance; and to members of *Trikone* Michigan for making us welcome in Ann Arbor. Special thanks to Ashok Row Kavi for our email conversations over the years, from which I have learnt so much, and for his never-failing support of my work.

I thank Kirti Singh, Y.P. Narula, Tara, and Namrata for their affectionate hospitality on our visits to Delhi; Shohini Ghosh for useful comments on some draft chapters, for references to important sources, and for all that I have learnt from her in years of chatting about gay topics and responding to each other's work; Sanju Mahaley for various pieces of information, especially on Hindi movies; Suresh M.S. for collecting and sending many news clippings and references; Prabha Dixit and Archana Varma for enlightening conversations; and Sujata Raghubir for her loving support.

I am more grateful than I can say to Saleem Kidwai, for getting me started on learning Urdu, checking my translations of *rekhti* poetry, locating sources for me, reading and commenting in detail on the entire manuscript, and for years of discussion, in person, on email, and on the phone, on everything from poetry to cinema to life and love. His input has greatly enriched this book.

Thanks to my parents, who spent the fellowship year with us in Ann Arbor, thereby blessing our home and work, and to my Uncle Rupin for his inspiring example. Our cockapoo pup, Pyara, who grew with this book, made my work much more enjoyable by his constant presence under my desk.

Finally, no words can fully acknowledge my partner Mona Bachmann's contribution—she cheerfully took on many chores so that I could work undisturbed, helped me procure textual and other sources, sat through numerous Hindi films, drove me to interviews in other towns with Hindi film songs playing repetitively, and interrupted her own work to read and comment in detail on several drafts of the manuscript; her comments have vastly improved the final product, although any mistakes that remain are my own. Without her presence in my life, this book would literally not have been possible.

[illegible] of discussion [illegible] on the phone [illegible] everything from poetry to [illegible] life and love. His input has greatly enriched this book.

Thanks to my parents, who spent the fellowship year with us in Ann Arbor, [illegible] and to my uncle [illegible] for his inspiring example [illegible] with this book, made my work [illegible] by his constant presence [illegible]

Finally [illegible] acknowledge my partner, Mona [illegible] [illegible] text [illegible] interviews [illegible] [illegible] through any [illegible] [illegible] like this book [illegible]

Chapter 1

Introduction

Marriage is a union of two spirits, and the spirit is not male or female.

—*Hindu priest, in conversation with me, 2002*

'Put a human face on it. Let's not talk about it in theory. Give me a story. Give me lives.'

—*Gavin Newsom, mayor of San Francisco, after he authorized same-sex marriages in the city, 2004*[1]

This book demonstrates that same-sex marriage, a social reality in India as well as the West today, has roots in the past. Its antecedents—marriage-like unions—appear in a variety of Euro-American and Indian texts. I argue that both in the past and in the present, mutual consent and family and community recognition validate a marriage; and this extra-legal validation sometimes extends to same-sex marriages too. While state recognition is desirable, the state's refusal to recognize a union as marriage does not mean that the union is not a marriage.

This book arises from three moments. One was in June 1980, when I, then living in New Delhi, the capital of India, read newspaper reports of a joint suicide attempt by two women, Mallika, 20, and Lalithambika, 17, in Kerala, south India. They left behind letters stating that they could not bear their imminent separation. These reports were followed in subsequent decades by a series of reports of such joint suicides, and also reports of same-sex weddings in different parts of India. Of the weddings, the most widely reported was the marriage, by Hindu rites, of policewomen Leela Namdeo and Urmila Srivastava in 1987 (see photo 5.1, page 131).

The second moment was in 1996, when I first read a fourteenth-century sacred narrative about the birth of a heroic child, Bhagiratha, to two women who make love with divine blessing. In subsequent years, I discovered more versions of this narrative, and continue to ponder its meaning.

The third is the present moment in 2005, when struggles are being waged worldwide for the legalization of same-sex marriage. Two countries (the Netherlands and Belgium), several provinces in Canada, and the state of Massachusetts in the United States have legalized same-sex marriage. Canada and Spain are on the verge of doing so. Most European countries give same-sex couples some of the rights of married people.[2] In some, like France and the Scandinavian countries, these partnerships can be registered as civil unions. The U.K. is in the process of legalizing same-sex civil unions.

These events in vastly different cultures, times, and places, all point to the possibility of same-sex commitment being recognized. Death, parenthood, and marriage—each is a rite of passage, and each may also, in the right circumstances, become, in Shakespeare's words to his male beloved, the "perfect ceremony of love's rite."

What is Marriage?

Many kinds of marriage have been outlawed in different societies; among these are widow and divorcee remarriage, intercaste and interracial marriage, and same-sex marriage. Marriage has varied so widely over time and space that its only core component is commitment.

Commitment between two persons of the same sex is not inherently different from commitment between persons of different sexes. "Gay marriage" is a misnomer.[3] A marriage is not gay (though the two partners may define themselves as gay). Being gay is just one dimension of a person, and marriage encompasses the whole person. Most of the same-sex couples who married in India did not define themselves as gay. When people claim the right to marry, their sex or sexuality is not intrinsic to that right, although social prejudice makes it appear so.

If those who claim that marriage has always, in all civilizations, been "between a man and a woman" are right, then the idea that two people of the same sex might marry one another suddenly appeared in the late twentieth century, ex nihilo, out of nothing. But ideas do not appear out of nowhere. The Hebrew Bible remarks, "there is no new thing under the sun," (Ecclesiastes 1:9), and the Sanskrit epic, the *Mahabharata*, declares, "That which does not occur here occurs nowhere." (*Svargarohanika Parva* V; XII: 291).[4] These claims, if read metaphorically rather than literally, indicate that phenomena grow out of earlier phenomena. They may take new forms but are never entirely new.

Same-Sex Desire and Love in Earlier Societies

Same-sex sexual relationships have been attested in almost all societies that have left written texts. These include ancient and medieval India, ancient Greece, Rome, Egypt, and China, medieval Japan, Western Europe, and Persia, and several Native American and African tribal cultures.[5] Historian Greg Reed has written about the ancient Egyptian tomb of two men, manicurists to the king, which uses conjugal iconography to represent them in an eternal embrace.[6] John Boswell has uncovered evidence from Eastern and Western Europe of same-sex unions performed by medieval Churches, and also ancient Greco-Roman same-sex pairings.[7] Weddings between men took place in the Roman Empire. In ancient Greece, male lovers swore fidelity at the tomb of Heracles (Hercules) and his beloved Iolaus.

Historians have written about conjugal relations between people of the same sex in later societies too. Striking examples include Eleanor Butler and Sarah Ponsonby ("the Ladies of Llangollen") Irishwomen who eloped and lived together for fifty years (1778–1829) in Llangollen, Wales, where their home is still a tourist attraction, and late nineteenth-century British poets, Katherine Harris and Edith Bradley, who lived and wrote together under the joint pen name "Michael Field." These couples used marital language to express their feelings for each other. The ladies of Llangollen in their diaries refer to one another as "My Beloved" and "My Love," and friends called each the other's "better half." Harris and Bradley wrote love poems and letters, describing exchanges of vows, garlands, and rings, Harris addressing Bradley as "my Bride." Harris nursed Bradley till her death from cancer. Harris died nine months later of cancer developed from grief. Anne Lister (1791–1840), a British heiress whose diaries show that she had sexual liaisons with many women, exchanged rings with her beloved, Ann Walker, in 1834, in a union solemnized at the parish church.[8]

Same-sex sexual activity and pairing is attested in many species—a recent book by a biologist documents the evidence underplayed by most scientists.[9] Human cultures institutionalize sexual activity and emotional relationships in many forms, including marriage of different types, prostitution, and liaisons. All of these institutions have also been hospitable to same-sex relations.

Marriage: Taking the Longer View

Opponents of same-sex marriage in the West often refer to "three thousand" and even "five thousand" years of history and "civilization" as evidence for marriage having "always" been between a man and a woman. Given that the legal precedents they refer to in fact have a history of about two hundred years, looking at other cultures helps us take a longer view.

All laws originate in custom, that is, in the social practice of local communities. Customs change gradually so the change is not startling to people. But when laws are written down, the written law freezes while customs continue to change. It takes time for the written law to catch up with custom. When the written law changes, people panic because they think written law changes practice, whereas in fact, changes in practice precede changes in written law, for example, incompatible spouses used to separate and remarry long before divorce and remarriage became legal. Similarly, same-sex marriages occurred long before laws began to recognize them.

Marriage varies widely over time and space. Formerly important components, such as the prescription to marry within one's own ethnic and religious group, have now been eliminated from the law in democracies. Monogamy is now required both in Hindu and Judeo–Christian law, but was not required in ancient Hindu and Biblical traditions. Divorce used to be hard to obtain in orthodox Christian and Hindu communities. It is now legal in both communities. Nor do spouses always vow to spend their lives together. Muslim Shia law recognizes a type of marriage called Mutaa, legal in India, which exists for a period of time specified in the marriage contract, which may range from one day to several years.[10]

The requirement that marriage lead to procreation was closely linked to the early age of marriage in premodern societies. Shakespeare's Juliet was married at fourteen; traditional Hindu law required a girl to be married before puberty. These laws have now changed—people can marry only as adults, and, since contraception is widely available, this means they can choose whether or not to have children. More important, laws now allow people past the reproductive age to marry. So marriage is no longer inseparably linked to procreation.

Even the prohibition on incest is not consistent worldwide. Marriage between first cousins is legal in all European countries, but illegal in some states of the United States. Marriage between cousins and between maternal uncle and niece is acceptable in several communities in India. Recently, scientists have finally noticed that despite modern Americans' horror at the idea of cousin marriage, cousins do marry in most parts of the world, and did marry in the United States as well, until the early twentieth century.[11]

The Indian government defines incest for purposes of marriage according to whatever definition is customary in the community of either partner.

Married relationships in all societies require commitment, but not all require exclusivity. Several premodern Indian love stories tell of passionate love between a woman and an already-married man who has no intention of divorcing his earlier wives. Marital commitment exists within certain temporal limits but those limits vary widely—from the traditional Christian "till death do us part" to the Hindu marriage "for seven lifetimes," to marriages for some months or years, to contemporary marriages, imbued with the awareness of possible divorce and remarriage, an awareness often documented in prenuptial agreements.

Each time a legal change was made in the modern era (legalizing divorce, property rights for wives, interreligious and interracial marriage in the West; and legalizing widow remarriage, intercaste marriage, women's property rights, and banning child marriage in India) politicians, the media, and religious sects claimed that these changes would destroy marriage, the family, and civilization.[12] A similar panic is being created now in Western democracies, vis-à-vis the demand for same-sex marriage.

Something Old, Something New

To adapt a popular bridal jingle, marriage is a mix of something old and something new. It has borrowed from many cultures and traditions to emerge in its present incarnation, and it is still changing.

True, same-sex marriage in its contemporary form is a new thing, but so is cross-sex marriage. To a Biblical patriarch like Abraham, marriage between American men and women, as "defined and protected" by the U.S. government today, would be unrecognizable. Sarah was Abraham's half sister (they had the same father and different mothers)—on that ground alone, their marriage would be ruled illegal by the U.S. government, despite the fact that many U.S. officials constantly quote the Bible to shore up their definition of marriage.

If marriage in the United States today nevertheless has enough in common with Abraham's marriage to Sarah that both can be termed "marriage," then same-sex marriage too has enough in common with cross-sex marriage to be termed "marriage." I do not claim that same-sex unions in premodern societies were exactly like present-day same-sex marriages. I claim that they were unions that had enough in common with present-day marriages to be termed antecedents.

Among the antecedents of same-sex marriage that I consider are: (a) representations of same-sex marriage in premodern texts (b) representations of marriage-like lifelong unions in premodern texts (c) representations of sex change enabling same-sex marriage (d) representations of same-sex couples aspiring and yearning to be married (e) same-sex unions within conventional families (f) use of tropes of marriage to describe same-sex relationships.

Same-Sex Suicides and Weddings

Lalitha and Mallika attempted suicide in 1980; in December that year, two other women, Jyotsna and Jayshree, committed suicide by jumping in front of a train in Gujarat, western India.[13] They left behind a note, stating that they could not bear to live apart after their marriages (to men) a few months earlier. Subsequent years saw many more such reports, which I discuss in chapter 4. It was as if the attempted suicide of Lalithambika and Mallika and the wedding of Leela and Urmila alerted the press to a phenomenon they had not noticed earlier.

In almost all cases (of suicide and of marriage), the partners were women (there are a few reports about men). They were lower middle class or working class. Most lived in small towns; a few in rural areas or big cities. They had some education and several were employed, but they were not primarily English speaking. Several of the marriages were cross-caste but some were within the same caste. Most of the individuals were Hindus, but there were also a couple of Muslims and Christians. The couples who entered into these unions in the 1980s and 1990s were not connected with any LGBT (lesbian-gay-bisexual-transgender) movement. In the last couple of years, human rights and LGBT organizations have begun to come to their assistance.

Police often collude with families to harass female couples, accusing one of kidnapping the other. This is a strategy parents routinely use to harass cross-sex couples who elope. But whenever same-sex spouses have been produced in court, following family complaints to the police, Indian courts have affirmed the women's right to live together. For example, in December 2004, Raju, 24, a Dalit girl, and her childhood friend, Mala, 22, a Jat, eloped to Delhi and got married in a temple by Hindu rites. The families tried to get them arrested, but when they were produced in court, local magistrate Sanjeev Joshi said they "enjoyed the constitutional right to live according to their wishes and no one had the right to interfere in their lives."[14]

As opposed to the majority of homoerotically or bisexually inclined people in India today who engage in same-sex liaisons while married to an opposite-sex partner, these female couples, through their weddings or suicides, made their love visible in cultural terms that are widely understood. Sometimes a wedding-like event was followed by suicide; in other cases, the wedded couple was accepted by the family of one partner and lived in a joint family.

Over the decades, as I collected these reports, as the LGBT movement matured in India and abroad, and as I researched the history of same-sex relationships, my questions grew. Why did these women choose these particular actions, gestures, and words, and how were their choices heard and understood by their families and communities, by state and religious authorities, and by the media? How did these women initiate social change by demanding that family, religious, and state authorities respect their rights?

What's Love Got To Do With It?

Mallika and Lalithambika spoke a cultural (and cross-cultural) language of love. Without using any term referring to sexual identity or behavior, they spoke their love for one another in terms immediately understandable by anyone. The press reporters understood the message, and relayed it. At one level, the message is that of all socially illicit lovers from Romeo and Juliet or Heer and Ranjha to the numerous doomed lovers of Hindi movies. Lalitha's words in her note, "I cannot part with Mallika. . . . Bury us together" are conventional—their unconventionality arises only from the gender of the lovers.

Mallika's letter is defiant: "Lili, after all everybody knows about our love. So here are a thousand kisses for you in public." Paradoxically, this love is private yet "everybody" knows about it, which drives them to consummate it in that most private of acts—suicide. Yet, at the same time, the suicide is as public as a wedding—through their embrace in death and the notes they leave behind, they kiss in public (which even married couples rarely do in India). Through the joint burial they request, they aspire to marriage in and beyond death.

Reporter Victor Lenous, noticing that the diary and book the girls left behind had both their names written in them, remarks, "The few personal belongings recovered also reveal their desire to share everything."[15] That sentence gets to the heart of the matter—the girls desire each others' kisses, embraces, or bodies but they also desire "to share everything." This is a desire for a particular type of relationship, a relationship of complete sharing often called marriage.

"Sweeten My Imagination": Love and Sex

Homophobia today derives its impetus from an emotional antipathy to sexual acts labeled disgusting.[16] This over-emphasis arises from centuries of negative obsession with sexual acts in the Christian West. A catchall term for these acts was the over-determined "sodomy," which includes oral and anal sex, regardless of the gender or species of the participants.[17]

This inappropriate focus on sex results in reducing individuals to sexual beings and relationships to sexual acts. Thus, when the Episcopalian bishop Gene Robinson was consecrated in 2003 in the United States, an opponent stood up and read out a list of sexual acts, which, he said, are performed by homosexual men. He neglected to mention that homosexual men also perform acts of love, both sexual and nonsexual, not to mention acts of everyday domesticity such as grocery shopping, cooking, and dish-washing. The Roman Catholic Church's position that it hates the sin (homosexual acts) but loves the sinner (the individual who does those acts) makes sense only if sex and desire are artificially abstracted from their emotional and social contexts. One is tempted to remark that, like the mad Lear, those who over-focus on sex in this way need "a little civet to sweeten [their] imagination."

The debate on same-sex marriage has refocused public attention on loving relationships between persons of the same sex, that include but are not limited to sex acts. Sex acts can occur in nonemotional contexts or as expressions of violence, but they do frequently occur in loving relationships. Imaginative literature often expresses an aspiration to such relationship. For example, in the context of medieval urban cultures of male-male eroticism, sexual attraction between males is assumed as an experience most men have, but love between men also appears as a desideratum. In his Urdu narrative poem, "Advice to a Beloved," Najmuddin Shah Mubarak "Abru" (ca. 1683–1733) describes the cosmopolitan atmosphere of Delhi, where beautiful young men are much courted. He describes the allure of beauty but points out that beauty fades while love lasts, and that one should therefore seek and find a true lover: "Now may all lovers get their hearts' desire,/May those who yearn be blessed with love's sweet fire."[18]

Prescriptive Antipathy versus Imaginative Sympathy

With some exceptions, those premodern texts in Euro-America that represent same-sex love, including sexual love, tend to be narratives and lyrics that appeal to the reader's emotions and imagination, while those texts that document sexual acts tend to be prescriptive, such as legal and medical texts. These texts prohibiting *certain sexual acts* tend to obscure in the public imagination over two thousand years of European writing about *love* between men and between women.

A similar, though less powerful, opposition between prescriptive and narrative texts is found in the Indian corpus. I first became aware of it through the fourteenth-century devotional texts, in Sanskrit and in Bengali, that tell the story of the miraculous birth of a heroic child to two women. This story appears unique but is not entirely so. It derives the idea of a female-female sex act leading to conception from an ancient Sanskrit medical text, which prescribes that if two women have sex and one becomes pregnant, the child will be born without bones.

In chapter 5, I develop an argument about these texts that is in some ways at the heart of this book. I suggest that the devotional texts' transformation of the monstrous into the miraculous is enabled by the emotional context in which the sex act is placed. The medical texts are not concerned with emotion. They focus on the abstract question of a sexual act between any two women anywhere. But the devotional narratives are imbued with empathetic emotions evoked by specific characters in a specific context. Negative emotions of distaste and anxiety are not entirely absent from the representation of the two women's lovemaking. But the texts resolve these by framing them in the larger context of love, human, and divine.

The reader is encouraged to wonder at the miracles worked by the Gods, especially the God of love, Kama, who inspires these two women's relationship. A wonderful evocation of the bliss and serenity embodied in Kama is the eleventh-century statue from Orissa in eastern India, which shows him shooting an arrow at two women (see frontispiece).

An emotional response, such as anxiety and disgust, can be fully countered only by another emotional response, such as sympathy with love or justice. It cannot, in my view, be entirely countered by logic or a rational argument alone, important though these are. When the U.S. state of Vermont was debating same-sex marriage in the late 1990s, many senators who had initially opposed the measure changed their minds after listening to the testimony of couples who had lived together for years, and voted in favor of same-sex civil unions which are now legal in Vermont from 2000. In 2004, following the Massachusetts Supreme Court decision legalizing same-sex marriage, San Francisco mayor Gary Newsom, who is a heterosexual Catholic, instructed city officials to perform same-sex weddings. The first of 4,037 weddings that took place was that of Phyllis Lyon, 79, and Del Martin, 83, who had lived together for fifty-one years. This wedding was seen on television nationally and internationally, and had a lasting impact, even though six months later the California Supreme Court voided the weddings. Two hundred weddings performed in New Paltz, New York, were upheld by a judge and remain valid. The weddings and joint suicides reported by newspapers in India have a similar effect.

Modern Homophobia and the Law

The extreme homophobia we witness today, manifested in lynchings and murders of gay people in the West, public executions of them in the Middle East, and violence against gay people and calls to persecute them in many countries, including India and Nepal, is a product not of the ancient or medieval past, but rather of modernity.[19] Even though medieval European churchmen condemned certain same-sex sexual acts, these acts were rarely punished as crimes. Rather, they were treated as sins to be expiated by religious penance.

Furthermore, celebration of romantic same-sex relationships, especially between men, is documented in medieval literatures in Europe, West Asia, and South Asia, despite Christian and Islamic and occasional Hindu condemnations of certain sexual acts.

This changed with the Renaissance, which is generally seen as inaugurating modernity in the West. From the fourteenth century onwards, homosexual acts were redefined as crimes rather than sins, the business of the state rather than the church alone, and historians have documented a dramatic increase in state persecution of homosexually inclined people, as also of other deviants, such as those perceived to be witches.[20]

With the development of the nation-state and Euro-American ascendancy in the modern world, the drive to impose uniformity in sexual matters acquired a new force and urgency. The modern nation-state is more efficient in imposing uniformity, because of developments in transport, communications, information gathering, and policing. New types of censorship and self-censorship also appear in the modern era.

That modern homophobia is so intense in the United States is not surprising, since it is a modern state par excellence. It is a product of the

Renaissance, and its Puritan founders, many of them fleeing religious persecution in Europe, disowned their medieval heritage, and instituted their own virulent persecutions of witches, sodomites, and religious minorities. Native Americans were much more tolerant of same-sex relationships, but their cultures were steamrolled by that of the colonizers.

Although the principles of the Enlightenment, proclaimed by the founding fathers in the U.S. Constitution, did curb these practices, homosexual relations, unlike religious dissent, did not become legal in the eighteenth or nineteenth centuries. Homophobia remained enshrined in law, and significant numbers of those termed sodomites were legally tortured, mutilated, or executed until the nineteenth century, and fined, imprisoned, and stripped of their positions till the twentieth century in both Europe and the United States.

South Asia has no extended premodern history of persecuting people for same-sex relations; in our research for *Same-Sex Love in India* (2000), Kidwai and I did not find a single documented instance of execution for such relations. Under colonial rule, what was a minor strain of homophobia in Indian traditions became the dominant ideology. The British introduced in India, as in most countries they colonized, a law criminalizing any type of sex other than penile-vaginal penetrative sex. In India, this law, prohibiting sex "against the order of nature" (Section 377, Indian Penal Code, 1860) remains on the books and has generally been interpreted to refer to anal or oral sex between men or between a man and a woman.[21] It is widely used to blackmail and harass gay men, and provides the basis for governmental discrimination, such as the dismissal of army officers.[22] It has also been used to threaten female couples, examples of which appear later in this book. It is very similar to the anti-sodomy laws that were struck down by the U.S. Supreme Court in *Lawrence vs. Texas* in 2003. Ironically, the United Kingdom legalized sex between consenting adults in 1967.

In March 1994, following the Delhi prison authorities' citation of Section 377 to justify their refusal to make condoms available to male prisoners, the AIDS Bhedbhav Virodhi Andolan (AIDS Anti-Discrimination Movement), known as ABVA, filed a public interest litigation petition in the Delhi High Court, challenging Section 377 IPC as violative of the Constitutional rights to life, liberty, and nondiscrimination, and obstructive of AIDS prevention work. The case was admitted in 1995, but was dismissed in 2001 when it came up for arguments as ABVA,—the only major AIDS organization in India that is entirely non-funded and run by unpaid volunteers, and therefore has limited resources—was unaware it had come up, and failed to appear in court.

In 2001, following a police raid on the office of an anti-AIDS organization, Naz Foundation (India) Trust, Lawyers Collective (whose HIV-AIDS unit was set up in 1998) and the Trust jointly filed a petition in the Delhi High Court, asking that Section 377 be "read down" to apply only to sexual assault on children. This petition also pointed out that Section 377 violates the Constitutional rights to life, liberty, and nondiscrimination, and has a devastating impact on AIDS prevention work. But it asked for reinterpretation rather than deletion of Section 377. This second petition, admitted in 2003, received widespread media attention, and garnered support from human rights and women's rights organizations, and from celebrities in different fields. The Government of India defended the law, claiming, among other things, that Section 377 is used primarily to punish child abuse. In September 2004, the court dismissed the petition, on the grounds that the petitioners were not being prosecuted under Section 377 and therefore had no cause for action against it. This dismissal is now being challenged in the Supreme Court of India.

Modern homophobia is deeply intertwined with modern nationalism. Most Indian nationalists, who fought for Indian independence from British rule, including M.K. Gandhi, accepted the British rulers' view of homosexuality as a vice. Several Indian politicians and opinion makers defend Section 377 as "Indian," and decry homosexuality, which they declare a product of the West.

This is one dimension of a general antisex attitude Indians inherited from British Victorians. Historians have discussed how nineteenth-century Indian social reformers and nationalists excised the erotic in their translations and publications of premodern literature, or interpreted nonjudgmental portrayals as denunciations.[23] Kidwai and I examine the specific effects of this on homoerotic literature, especially the censorship of female homoeroticism in Urdu Rekhti poetry (see chapter 8), the heterosexualization of the Urdu *ghazal*, and modern translators' misinterpretation of the Sanskrit *Kamasutra* as a warning tract.[24] These trends paralleled developments in Victorian England, where novelists like Thackeray complained of new restrictions on depicting erotic life.

Modern homophobia, in India as in England, is most evident in the middle classes who shape educated public opinion, and less evident among the former aristocracy and the poor. Homophobia drives many Indians to lead double lives, flee the country, try to "cure" themselves, or even commit suicide.

However, modernity also produces new opportunities and aspirations. Increased urbanization allows for greater anonymity and membership in more than one community. Many people, especially women, acquire greater economic and social autonomy and mobility, and new aspirations to individual and collective equality and freedom.

Democracies claim to endow all citizens with equal rights and protections. This enables members of traditionally reviled groups to demand equality with other citizens. In 1967, the U.S. Supreme Court, in the case of *Loving vs. Virginia*, overturned laws banning interracial marriage, ruling that the marriage of Mildred Jeter, a black woman, and Richard Loving, a white man, was legal everywhere in the country, even though this type of marriage was traditionally outlawed and led to social violence. The Court declared that the right to marry is "one of the basic civil rights." It is this basic right that gay people are now seeking.

Civil Marriage, Religious Marriage, and Democracy

In secular democracies today, religious marriage is distinct from civil marriage. Governments recognize types of civil marriage that particular religious communities do not, for example, remarriage after divorce, which some churches, such as the Catholic Church, do not recognize. Many Jewish and Christian communities will not perform interreligious marriages, which the state validates. When my partner and I were planning our wedding, some rabbis who were willing to marry two women would not marry us because I am not a Jew. In the Netherlands and Belgium, the state recognizes same-sex marriages, but some churches do not.

Despite this apparently clear distinction, the dividing line is fuzzy in the minds of many. When President Bush accused judges of reducing marriage to a legal contract, thus cutting it off from its religious roots, he appeared unaware that civil marriage has always been a legal contract, cut off from religion. The secularizing of the Euro-American polity in the nineteenth century entailed the separation of Church and state, which compelled the state to recognize non-Christian marriages. This separation allowed secular democracies to legalize divorce in the twentieth century, despite Jesus' and St. Paul's unambiguous denunciation of divorce and many Churches' consequent refusal to grant divorces or remarry divorced people.

But some states, such as the United States, Denmark, Norway, Sweden, and Iceland, endowed religious officials, such as priests, ministers, and rabbis, with the power to act on the state's behalf and confer civil marriage rights on couples married by religious ceremony. In the United States, the partners must obtain a marriage license, but guests do not usually see it. They witness a religious marriage, which is also a civil marriage, and the differences are usually not clear to them.

The Commonwealth of Massachusetts
DEPARTMENT OF PUBLIC HEALTH
REGISTRY OF VITAL RECORDS AND STATISTICS
CERTIFICATE OF MARRIAGE

(State file number)
CAMBRIDGE
(City or town making return)
Registered No. 631
Intention No. 610

1 Place of Marriage
City or Town CAMBRIDGE
(Do not enter name of village or section of city or town)
2 Date of Marriage JUNE 11, 2004
(Month) (Day) (Year)

3 FULL NAME PARTY A: GEETA H. PATEL	11 FULL NAME PARTY B: KATHLEEN M. WESTON
3A SURNAME AFTER MARRIAGE: PATEL	11A SURNAME AFTER MARRIAGE: WESTON
4 DATE OF BIRTH: January 16, 1957 — 5 OCCUPATION: PROFESSOR	12 DATE OF BIRTH: November 2, 1958 — 13 OCCUPATION: PROFESSOR
6 RESIDENCE NO. & ST. 43 KENWOOD ST. CITY/TOWN BOSTON ST MA ZIP CODE 02124	14 RESIDENCE NO. & ST. 43 KENWOOD ST. CITY/TOWN BOSTON ST MA ZIP CODE 02124
7 NUMBER OF MARRIAGE (1st, 2nd, 3rd, etc.) 1ST — 7A WIDOWED OR DIVORCED	15 NUMBER OF MARRIAGE (1st, 2nd, 3rd, etc.) 1ST — 15A WIDOWED OR DIVORCED
8 BIRTHPLACE MUMBAI (City or town) INDIA (State or country)	16 BIRTHPLACE OAK PARK (City or town) IL (State or country)
9 NAME OF MOTHER/PARENT SHAKUNTALA NATH	17 NAME OF MOTHER/PARENT DARLENE WEINGAND
10 NAME OF FATHER/PARENT HIRALAL N. PATEL	18 NAME OF FATHER/PARENT WAYNE A. WESTON

19 THE INTENTION OF MARRIAGE by the above-mentioned persons was duly entered by me in the records of the Community of CAMBRIDGE (Name of Community) according to law, this 19 day of May 20 04
☐ COURT WAIVER ☐ AGE ORDER Issued May 25, 2004 (Month) (Day) (Year) by [illegible signature] (City or Town Clerk or Registrar)

20 I HEREBY CERTIFY that I solemnized the marriage of the above-named persons at No. 197 Auburn St.
(If marriage was solemnized in a church, give its NAME instead of street and number)
Cambridge (Name of city or town) on June 11 2004 (Month) (Day) (Year)
Signature [illegible signature] Clergy (Member of the Clergy, Priest, Rabbi, Imam, or Justice of the Peace, etc.)
Brigid [illegible] Frizell (Print or type name)
Address 197 Auburn ST Camb MA 02139

21 Certificate recorded by city or town clerk JUNE 21, 2004 (Month) (Day) (Year) [illegible signature] CLERK OR REGISTRAR

22 PARTY A SEX: ☐MALE ☒FEMALE	23 PARTY B SEX: ☐MALE ☒FEMALE

JUN 28 2004
A TRUE COPY ATTEST
[illegible signature]
City Clerk

Figure 1.1 Geeta and Kath's marriage certificate, issued by the state of Massachusetts. See also pages 236–7.

This overlap between the secular and religious functions of the same officials creates conflicts. Two Unitarian ministers who performed same-sex marriages in New Paltz, New York, in February 2004, and declared that they were doing so as civil officials, were prosecuted.[25] As ministers of their church they were entitled to perform same-sex marriages, but as civil functionaries they were not so empowered. However, when they perform the wedding of a heterosexual couple, the marriage is simultaneously both religious and civil.

The absurdity of this hairsplitting is most evident in a country like France where same-sex couples can engage in civil unions but not marriages. France recognizes only civil, not religious, marriages, but a French civil official who insisted on performing a same-sex union as a marriage, not a civil union, was prosecuted for doing so.

In chapter 2, I briefly trace the evolution of marriage in Euro-America and India from its origins as an arrangement by mutual consent between two individuals and families, to a sacrament controlled by religious organizations, to a contract controlled by the state. All three forms of marriage, I argue, are still alive, coexisting simultaneously and impacting one another.

Despite the interventionist and authoritarian character of the Indian state, Indian marriage law still allows for more variations in marriage, granting greater freedom to individuals and communities to validate marriages without state intervention. The degree to which people retain control over definitions of marriage varies from country to country in the West. Although the modern state increasingly controls all aspects of people's lives, democratic states, insofar as they remain democratic, cannot absolutely control marriage.

This is because democracies are in theory committed to individual liberty and equality, even though in practice democratic governments constantly try to infringe on these rights. Sree Nandu, a young woman in Kerala who recently asserted her right to live with her partner Sheela, in the face of family and police hostility, summed up the connection between same-sex marriage and democracy: "We made it clear to the police that we are majors, and the Constitution of India has given us the right to live life the way we want"[26] (see photo 4.1, page 93).

After a magistrate confirmed Raju and Mala's right to live together following their elopement and Hindu wedding ceremony, Raju was reported as saying they would fight for a change in law to give same-sex couples the same matrimonial rights as cross-sex couples.[27] This statement by a working class, Dalit young woman is a portent of things to come.

Because democratic Constitutions endorse citizens' equal right to life, liberty, and happiness, right-wing forces in the United States are afraid of the Constitution and anxious to write discrimination into it. They fear that the Constitution will be invoked, as Constitutions of several other democracies have been, to protect citizens' right to a same-sex union. For instance, in 2001, Germany passed a law allowing same-sex partnerships to be registered. It was challenged before the Supreme Court, because the German Constitution contains the provision, "marriage and family enjoy the special protection of the state." The German Supreme Court ruled in

2002 that allowing same-sex partnerships to be registered does not lower the state's protection of family and marriage.

Why Seek State Recognition?

If one's community recognizes one's religious or other wedding ceremony, why seek state recognition and the right to civil marriage? The answer is simple—because the modern state confers many privileges and benefits on couples it recognizes as married, and withholds them from long-term couples it does not recognize as married. In the United States it is estimated that married couples acquire 1,049 such rights, some of which may be crucial to their ability to live together and protect their resources.

The Indian government, since it does not dispense welfare and social security benefits, gives few privileges to married couples, but there are some significant ones, and many more if one partner has a government job. Indian employers, social welfare and charitable organizations, and institutions like life insurance companies also give many benefits to married couples, which they do not give to unmarried couples.

One female couple in Kerala tried to take advantage of such a privilege. Mini, 29, and Sisha, 19, worked together in an industrial unit for unmarried women. The employer gives Rs. 20,000 to any employee in this unit who gets married. In 2000, Mini agreed to a matrimonial alliance arranged by her parents, and took the money. Then she cut her hair short, put on male attire, and eloped with Sisha to Coimbatore in a neighboring state, where she changed her name to Babu and took a job as a man in a textile factory. The families filed a police complaint, so the women had to return four months later. But the court ruled that they could live wherever they wanted to, and they are determined to live together.[28]

Nevertheless, in India, family and community still confer more of the benefits of marriage than does the state, so family and community approval is sought and valued more than government approval. This partly explains why several of the female-female weddings that have occurred recently with family support, have been accepted by the local community, without recourse to the state.

Likeness and Difference

The debate about same-sex marriage is largely a debate about sameness and difference. Those opposed to same-sex marriage claim that it can never be like the "real thing," which is marriage between a man and woman. Some gay activists insist that gay people should retain their positive differences from heterosexual people, which they are in danger of losing if they get married.

My assumption throughout this book is that for purposes of marriage, same-sex couples and cross-sex couples are more alike than different.[29] Whether or not a long-term committed loving union in which partners share financial resources and are each other's primary caregivers is termed marriage or not makes little difference. A long-term couple generally has certain irreducible needs, for example, the need for access to one another, and the need to make certain joint decisions.

It is also a mistake to assume, as some gay activists do, that all heterosexual marriages are conventional while same-sex marriages are unconventional. Heterosexual marriages that appear conventional may not be so—celibate marriages, partner-swapping, nonmonogamous marriages, and couples living in separate residences all occur more than one may realize, even in India. Conventional marriages may also be more flexible, compassionate, and dynamic than we give them credit for. In chapter 7, for example, I discuss the ways in which highly unconventional arrangements flourish in the bosom of the supposedly conventional family—witness the example of a lower middle class man in a small town who agreed, on his wife's insistence, to marry her female lover.

Conversely, many same-sex couples prefer to be conventional—monogamous, sharing all resources, even having one partner be a full-time homemaker. While cross-sex couples have options available to them, same-sex couples do not.

Some leftwing activists argue that state-conferred privileges of marriage should be abolished—people can marry by religious ceremony, but the state should stay out of the arena of marriage. However, there is no campaign anywhere in the world for abolition of civil marriage, so discussion of that possibility is purely academic, and has no effect in the real world. Instead, human rights defenders are campaigning today to include those couples who are excluded from civil marriage.

Is Marriage Always and Only Oppressive?

In her pioneering study of chosen kinship, sociologist Kath Weston notes, "Change and continuity are more closely related than many people tend to think. No search is more fruitless than the one that seeks revolutionary forms of social relations which remain 'uncontaminated' by existing social conditions."[30] Several feminist and queer theorists and activists in the

United States argue that marriage is a heterosexist, patriarchal institution not worth recovering because its history is fraught with the oppression of women and children. This assumes that marriage has always and everywhere been oppressive, and continues to be so. It also assumes that all lovers, cross-sex or same-sex, who aspire to marry are deluded by what orthodox Marxists term false consciousness.

One problem with this argument is that it does not probe deep enough. If the oppressiveness of marriage is based largely on male domination, would that domination disappear if marriage as an institution were abolished? In societies like the United States, where many people live together without marriage, the incidence of boyfriends beating and even killing girlfriends is as high as that of husbands abusing wives. The reduction of male oppression requires empowerment of women and children. When that empowerment occurs, as in some Scandinavian countries, marriage becomes more egalitarian.

Abolishing marriage will not dismantle patriarchy or heterosexism, but institutional empowerment of women and gay people is a move toward such dismantling. Legalizing same-sex marriage involves the institutional empowerment of gay people. It allows, for example, a person opting out of or thrown out of a same-sex union, or a battered partner to claim the rights of divorce, alimony or maintenance, custody or visitation rights vis-à-vis children, and social recognition of loss. As of now, such a person has to deal with his or her grief without social or state support.

The second problem with advocating a boycott of marriage is that it is unrealistic. If we are to opt out of all patriarchal institutions, we must refuse school and university education, boycott voting and public office, shun government and corporate employment, and neither buy nor sell goods. Schools, universities, governments, corporate business, and the market, as we know them, are all patriarchal institutions. Such a program would lead to extreme separatism, practiced by some small religious sects and left-wing or right-wing communes. Since few people take these routes, the institutions concerned would be unaffected.

Third and most important, while marriage has enabled and continues to enable a great deal of oppression, it has also aspired toward and sometimes embodied friendship and love. Happiness is notoriously hard to measure, so without making any grand claims with regard to happiness, it is important to remember that many married people in most societies claim to be and are seen as being happy, a fact that helps inspire young people to marry. Excluding blatantly violent and oppressive marriages, there is no evidence that most unmarried people are happier than most married ones, or that most unhappy married people would be happier if they were single.

A few theorists also argue that not just marriage but monogamous coupling itself is oppressive because it restricts sexual freedom and marginalizes single, celibate, and promiscuous people. These theorists overlook the fact that many people make a conscious choice to restrict their sexual freedom to gain other freedoms and pleasures available within monogamy. This is not surprising because humans give up certain freedoms and pleasures, such as the pleasure of uninterrupted leisure, in order to gain other pleasures, such as the pleasure of productive work.

Any democracy that protects the freedom to marry should equally protect the freedom to be single, celibate, or promiscuous. So far, in known history, even in democracies that protect all of these freedoms, most people still choose to live most of their adult lives coupled rather than single. Theorists are free to view this choice as unfortunate or restrictive, but that is no reason to deny gay people the right to marry.

Power and Equality

While same-sex couples do not face the challenge of gender inequality that cross-sex couples do, this does not remove power, inequality, or hierarchy from same-sex relationships or from the families gay people build. If Foucault has taught us nothing else, he should have taught us that totally eradicating power is impossible. The quest for absolute equality might perhaps be fulfilled in a world of clones, although even there it might face some obstacles contingent on age and experience. Nor, in my view, is all hierarchy or inequality inherently negative. Inequalities or differences in ability and experience, provided they are not reified or assigned on the basis of biological category, can usefully complement one another in relationships, and can also foster mutual learning.

Awareness of the constantly shifting power balance in any relationship, along with strong filiations to others outside the relationship, and equitable sharing of resources in forms that are backed up by society and state, should work to check abuses of power within relationships. The demand for state recognition of same-sex marriages has arisen in the context of movements for social, economic and political equality, and freedom, and of major socioeconomic changes that made these movements possible.

Democracies already recognize many more types of families than they did a century ago. Single mothers in the West are no longer social outcasts as they were just a century ago. Many children have only one parent; many others have more than one parent of each gender, following their biological parents' divorce and remarriage to others. The "family values" that have become a mantra in the United States can no longer be about valuing a family consisting of one father, one mother, and their biological children, because few families conform to that rigid model.

In India, single mothers are still heavily stigmatized. Widows or divorcees with children rarely remarry, so relatively few families comprise "my children," "your children," and "our children," but such families are beginning to emerge. Major changes have occurred in adoption patterns. Adoption of children from orphanages is now widespread in the Indian middle class, in contrast to the traditional practice of adopting a relative's child. I know several single women and some single men who have adopted children, and also several same-sex couples who have raised children together in India.

Ideals of Modern Marriage: Derived from Ideals of Same-Sex Friendship

The distinction between marriage and friendship was not always absolute. In one long-term male-male relationship in medieval India, poet Maulvi Mukarram Baksh's companion Mukarram inherited his property. Mukarram observed *iddat* for the Maulvi. *Iddat* is the period of sexual abstinence and mourning a widow observes for her husband.[31]

I claim that modern male-female marriage, which, in today's democracies, has shed most of its legally oppressive features in its aspiration toward friendship, acquired its ideals of friendship not from a heterosexual model but from the model of same-sex friendship (see chapter 5). In different cultures, these ideals include some of the following: friendship is central to human happiness; friends desire to live together and share everything; friends have all their property in common; a friend is a second self or a better half; one is spontaneously attracted to a person destined to be one's friend; physical intimacy is an aspect of friendship but not the most important aspect; friends will sacrifice everything, even life, for each other; and friends desire to die together.

Readers today may find these ideas odd when applied to same-sex friends, but if the word "spouse" replaces the word "friend," they would recognize them immediately. However, for many centuries before the twentieth, texts in Western Europe, West Asia, and the Indian subcontinent projected these ideals as pertaining to same-sex friends. Same-sex friendship is an important ancestor of modern marriage. This is because, in most premodern societies, cross-sex marriages were usually family arranged while same-sex friendships were usually self-chosen. Modern marriage, based as it is on mutual choice, is closer to premodern friendship in that respect.

The most striking example of this influence of friendship ideals on marriage ideals is in the area of possessions. In his essay on friendship, seventeenth-century French philosopher Montaigne points out that these ideals of shared matrimonial possessions originate in the ancient ideal of male friends choosing to share everything: "That is why the lawmakers, to honor marriage with some imaginary resemblance to this divine union [of friendship], forbid gifts between husband and wife, wishing thus to imply that everything should belong to each of them and that they have nothing to divide and split up between them."[32]

In Euro-America as in India, the ideal that husband and wife own everything in common was not generally reflected in law. The husband controlled property, and the wife was rendered destitute if she left him or if he lost the property. Parents tried to protect their daughters through marriage settlements negotiated at the time of marriage. The idea that spouses' property is pooled and becomes joint matrimonial property, which is equally divided at divorce, became law in the West only in the twentieth century, after the concept of no-fault divorce developed. In India, the law is still murky on this issue, and divorce usually leads to bitter disputes regarding property division.

Today, it is taken for granted at least among the middle classes that husbands and wives should share property and finances, which leads governments to bestow joint financial benefits on them. But earlier Indian and Western ideals of friendship assume that one's close same-sex friend is one's second self and therefore shares all one's worldly goods.

It is ironic that some self-styled defenders of marriage today want to exclude same-sex couples from a modern institution whose ideals draw on earlier ideals of same-sex love. Just as irrational antiwomen prejudice prevented most premodern thinkers from conceiving women as capable of the highest type of friendship and love, so also at present irrational antihomosexual prejudice prevents many modern people from conceiving what the ancient Greeks could easily understand—that two people of the same sex may grow from youthful passion into mature union.

Love and Marriage: Not Quite a Horse and Carriage

The prevailing myth in the West today is that all marriages are or should be based on romantic love. Because of this myth, most Europeans and

Americans tend to represent their self-arranged marriages to others and to themselves as romantic unions.[33] However, this myth is less than two centuries old, and it coexists with a much older social suspicion that romantic love is indistinguishable from infatuation, and is therefore not enough to base a lasting marriage on.

This suspicion of erotic love is much more visible in India today than in the West. Marriage is near universal in India; most marriages are family-arranged, and love is expected to develop after marriage. The Indian English term "love marriage" indicates that marriages based on individual choice are still a minority of all marriages. Indians who oppose love marriages do so on the basis that love is really a cover for desire, therefore marriages based on it are unlikely to last, while family-arranged marriages, based on cultural compatibility, are more likely to last.

Both in India and the West, families often disapprove of cross-sex marriages on a variety of grounds, and not just the obvious ones like differences in class, caste, or race. For example, a Hindi movie *Dil Chahta Hai* (The Heart Desires, 1999) shows a son "coming out" to his mother about his love for a divorced woman, much older than he is. His mother encourages him to confide in her, saying she will understand, but when he hesitantly does so, she calls him mad, and berates him for disgracing her and ruining his life.

Unlike most of their counterparts in the West today, Indian heterosexuals who enter into or approve of love marriages often face social opposition, which can range from mild criticism to intensely violent hostility. It is important to remember that many male-female couples also commit suicide together when their families oppose their marriages (see chapter 4).

This situation allows for the possibility of an alliance in India between heterosexuals who approve of love marriage and homosexuals. All same-sex marriages are by definition love marriages while most cross-sex marriages are family-arranged. To those Indians who think the best marriages are family-arranged, same-sex love, since it is based on individual choice, seems less than legitimate. On the other hand, those who think that individual choice, love, and/or destiny play a role in marriage may be more likely to concede that same-sex love and marriage are plausible. Although still in a minority, they constitute a large and growing body of educated public opinion.

Arranged Love and the Family

Love marriages and arranged marriages in India today are not always polar opposites. Many marriages are an amalgam of both, as some type of individual choice is increasingly incorporated into family arranged marriages. Instead of merely seeing a photograph and meeting for the first time at the wedding, as they did a few decades ago, many couples now meet at least once before the wedding, and often have the right of refusal if they take a strong dislike to each other. Individual and family negotiate the degree of choice, which, for a woman, is often a mere simulacrum.

Often, a man and a woman who meet reluctantly under family auspices decide that they like or even love one another. In the urban middle classes, a family-sponsored engagement may involve premarital correspondence, exchanging birthday and Valentine gifts, or going out together, a process that fosters what may be called "arranged love." Arranged love has precedents in ancient Indian texts, where two people fall in love when their beauty is praised to each other or they see each other's pictures. In the Hindi movie *Hum Aapke Hain Kaun* (Who Am I To You? 1994), this type of marriage symbolizes Indian aspirations, when the hero, Prem (literally, love) asked if he wants an arranged or a love marriage, says he wants an "arranged love marriage."

A cross-sex love affair of which families disapprove does not always end in elopement. When both man and woman are educated and employed, conflicts are often resolved when the families realize the partners are not dissuadable. They then agree to make wedding arrangements and proceed to match horoscopes, give a dowry and follow all the rituals very much as if they had arranged the match in the first place. This process of spouses' parents accepting a fait accompli is found in ancient Sanskrit narratives, and also described in the *Kamasutra* (III. 5. 1–21).

These patterns help explain how some Indian parents today negotiate their children's choice of same-sex partners. Families participated in several of the female-female weddings reported in newspapers, and, when questioned, described how they tried to separate the couple, but gave in when they found their daughters adamant. Tanuja Chouhan and Jaya Verma, both 25, who married in April 2001 in Bihar, are nurses and met at work. Family and friends participated in their wedding by Hindu rites.[34] Even when couples elope and have a clandestine wedding in a temple, families may later arrange a wedding ceremony at home. This happened in the case of Leela and Urmila.

But love stories, cross-sex or same-sex, do not always have happy endings. Some stories in the eleventh-century *Kathasaritsagara* show parents accepting love matches and even elopements; others show them cursing the couple and the curse taking effect. Numerous Hindi films depict parents separating, confining, and threatening lovers, subjecting them to emotional blackmail, and forcibly marrying them to others. Some families torture and kill young people who elope.[35]

More commonly, families distance themselves from the couple. Social embarrassment is central to this type of reaction. On July 9, 1993, Meenu Sharma and Neeru Sharma were married by Hindu rites in a temple in Faridabad in north India. The marriage took place without their families' participation, and the witnesses were a dozen youthful friends. The families did not disown them but reported feeling socially disgraced. Meenu's sister said she was teased so much that she had to stop going to school.[36]

Even couples who obtain family approval face other kinds of social prejudice and discrimination. Nurses Jaya Verma and Tanuja Chouhan were criticized and abused by some neighbors. The landlord of their rental apartment told them to vacate the place. They said they would not let this type of harassment prevent them from living together.[37]

Police Violence Against Lovers

Instead of protecting citizens' right, police routinely collude with parents to deprive couples of their rights. The police actively help communities to inflict violence on male-female couples who elope. This same strategy is also used against female couples. For example, in April 2005, Allahabad police arrested two girls, Shilpi Gupta and Usha Yadav, who had eloped from Allahabad to Gujarat and had lived together for six months. Shilpi's parents accused Usha of kidnapping their daughter. The women declared before the police that they would commit suicide rather than be separated.[38] Shilpi's father admitted that she had repeatedly turned down proposals from men because of her lesbian relationship with Usha.

In India,which prides itself on being the largest democracy in the world, policemen and policewomen illegally intimidate young women and force them to separate. Sheela and Sree Nandu, a couple who themselves braved violent opposition (see page 93) and have now formed a women's shelter called Snehapoorvam (Unique Affection) in Thiruvanthapuram, Kerala, reported that two women, Rashiath, aged 20, and Neethu, aged 18, had sought shelter with them. Their parents were invited for discussion, and the women said they did not want to go home but wanted to stay together. While they were talking, policewomen appeared and forcibly took the girls to the police station. When Sheela and Sree Nandu protested, the police beat them. At the police station, many relatives and neighbors gathered, with police compliance, and blocked activists' access to the girls. The girls were produced before a magistrate, who upheld their right to live together. Yet, the police handed the girls to their parents, who took them home and forced them to sign papers saying they were willingly going home. They went back because police denied them access to any other support. Thus, the police constantly flout the rule of law and assert the violent rule of families over young adults, especially young women.[39]

Interreligious and cross-caste love affairs often trigger equally or more virulent reactions than do same-sex affairs. Not so long ago, black–white and Jewish–Christian marriages set off the same sort of violent reactions in the United States Interracial, international, interregional, interreligious, cross-class, cross-generational, and cross-caste marriages often face some degree of social hostility both in the West and in India. All these types of marriage privilege erotic love over social suitability.

Fear of Erotic Love

My argument is that the debate about same-sex marriage, both in India and the West, is really a debate about the value of erotic love (unlinked to procreation). In same-sex marriage, society confronts its deepest fears about the dangers of erotic love. Hence the paranoid questions constantly voiced by opponents of same-sex marriage in the United States: if same-sex marriage is allowed today, can incestuous, polygamous, and bestial marriages be far behind? These questions betray the fear that erotic love is a force that can run amok.

To borrow philosopher Martha Nussbaum's terms from another context, erotic love tends to be viewed as socially and morally inappropriate insofar as lovers' extreme value for one another is private, not public, and no one else can participate in its excess.[40] Perhaps the strongest component of many cultures' ambivalence toward erotic pleasure is the fear that it makes individuals selfish and indifferent to the welfare of others. Marriage is supposed to redirect this selfish passion toward the care of others—not only children, but also elders, relatives and friends, guests, and the poor. In Jewish wedding contracts, the spouses promise to make their home a hospitable place for guests. Hindu wedding ceremonies emphasize the couple's commitment to foster the welfare of family, community, and even animals. Invoking God or the Gods is a way to embrace concern for community, humanity, ancestry and posterity, which are larger than the couple's desire for each other.

Most wedding ceremonies incorporate parents and relatives, which may reinforce patriliny but may also integrate the couple into society's past and future. As I argue in chapters 5 and 6, the emphasis of most cultures on procreation, much in excess of any need for a larger population, is geared toward restricting the couple's indulgence in sexual pleasure, and redirecting their energies toward the supposedly unselfish task of raising children.

Even as society permits sexual activity in marriage, it also tries to constrain that activity, and since it is hard to police the conjugal bed, an insistence on procreation works as the best form of constraint. The child is the guarantee of the unselfishness spouses are supposed to offer their families and society.

In the West, suspicion of sexual activity stems in part from the ancient Greco-Roman view of erotic love as a socially disruptive madness, and in part from Christianity's ambivalence regarding all sexual love, including conjugal love. The ancient Greeks prayed to Aphrodite, Goddess of erotic love, to stay far from them. This pre-Christian ambivalence was reinforced in medieval Christendom by the idea that amorous love entices one to the sin of idolatry (preferring a human being to God), and to the sin of lust (preferring fleshly to spiritual pleasures).

Christianity, like Islam, was and still is conflicted about erotic love and romantic friendship—on the one hand, they can be seen as mirroring and leading to the love and friendship of God. On the other hand, they can function as rivals to the love of God. They lead individuals, like Romeo and Juliet, or Laila and Majnun, to worship one another, defy family and society, and become social outcasts.

In the context of this ambivalence toward desire, same-sex relationships, being biologically infertile, are tainted with the suspicion of being entirely selfish, based on sexual pleasure alone. Since everyone knows that the intensity of sexual desire does not last, they are also seen as doomed to transience. It is for these reasons, among others, that right-wing forces view them as unworthy of admission to the institution of marriage.

Love, Suffering, and the Larger Good

Love is seen as unquestionably good only when it enables the welfare of others besides the two partners. This is true even in the West today, despite the new emphasis on sexual fulfillment as a goal of life. In the defining myths, lovers must suffer and make sacrifices. When these sacrifices are only for one another, couples often end up dead—Romeo and Juliet, Tristan and Isolde, Laila and Majnun.

When lovers are perceived as struggling for justice or for a cause larger than their own desires they become heroes. Ancient Greek historian Thucydides recounts that Athens publicly honored sixth-century BC male lovers Harmodios and Aristogeiton for having overthrown the tyrant Hippias, whose brother Hipparchos had tried to seduce Harmodios. Stories of suffering lovers fueled the battle to legalize interracial marriage in the United States.

In Christendom, where the ultimate symbol of love is a painful crucifixion, and where the seed of the church is said to have been watered by the blood of the martyrs, it makes sense that all love must prove itself as good through suffering and sacrifice. The word "passion," now connected with erotic love, means "suffering," and to have "compassion" is to suffer with someone.[41]

In India, love often has to prove itself through tests. When Mahatma Gandhi's son and fellow nationalist Rajagopalachari's daughter fell in love and wanted to marry, the parents imposed a period of separation to test whether their love was true or merely a temporary attraction. The idea that lovers must undergo ordeals is central to Indian cinema, and same-sex lovers are no exception. The 1998 film *Fire* that caused a huge controversy in India because of its lesbian theme, follows this convention. Both women undergo many ordeals, climaxing in a trial by fire. They are reunited only after they pass all these tests.

The ideal of sacrifice is not undiscriminating, though. Some sacrifices are seen as appropriate and others as inappropriate. In general, sacrifice is appropriate when made for a worthy person who would reciprocate, and who would refuse the sacrifice if s/he knew of it. Indian fiction and cinema rarely endorse the sacrifice of romantic love to parents' arbitrary prejudice. This is clear in the paradigmatic movie romance, *Devdas*, based on Sharat Chandra Chatterjee's Bengali novel, and filmed nine times between 1928 and 2003, in Hindi, Bengali, and Telugu. While both heroines defy social convention, the hero is unable to stand up to his bullying father and brother, thereby ruining his own and his beloved's lives.

Nor is sacrifice an ideal only for women; it is equally normative for men. Men are expected to sacrifice individual desires for family and community; this often takes the form of using their income to support parents, siblings, and other relatives.

In a set of interviews with gay male Christian couples in the United States, Paul, a musician in his late fifties, who has been in a monogamous relationship with another man for six years, comments on monogamy: "Christianity has something to do with this. . . . That's what love is about. It's about putting another person first. Regarding sex as just about having fun and enjoying yourself actually removes the possibility of sex being about something that you give. It makes it something you take."[42]

Newspapers in the West today report the sufferings of gay people who are not allowed to visit their long-time partners in hospital or attend their funerals. This type of story appeals not only to the reader's sense of justice but also to sympathy for suffering love. Stories about gay people who lost partners in the September 11 attacks were particularly effective, because they connected with a national sense of loss and heroism.

Same-Sex Relationships and the Social Good

Do sacred texts or traditions ever show same-sex relationships conducing to the social good?

The Bible and the Qur'an contain what appear to be prohibitions of same-sex sexual intercourse. These prohibitions have been interpreted in many different ways, and are not absolute condemnations, as literalists believe them to be.[43] However, they are not countered within these texts by an affirmation of same-sex sexual relationships. While the Bible celebrates passionate same-sex relationships, such as those of David and Jonathan or Ruth and Naomi, it does not represent them as sexual. Outside of the Bible and the Qur'an, though, Christian and Islamic literary traditions often celebrate loving same-sex relationships. The relationship of Emperor Mahmud and his beloved slave Ayaz is an example (see chapter 5).

Unlike Christianity and Islam, Hinduism has not one but thousands of sacred texts. Overall, Hindu texts, I argue, distinguish desirable relationships from undesirable ones, not solely on the basis of the partners' gender or the sexual acts performed but on the basis of how these relationships contribute to the greater good.

Relationships based only on individual pleasure are judged undesirable and shown to logically culminate in disaster, since two individuals who selfishly desire one another for their own pleasure may also desert each other when they discover that they can get greater pleasure elsewhere. Such are the many Indian language folk tales that show lovers eloping together only to discard one another for other lovers.

The most significant difference in this regard between premodern Indian and European texts concerns sexual intercourse. Alan Bray, in his book *The Friend*, has shown that long-term same-sex relationships were celebrated in public spaces from the fourteenth to the nineteenth centuries in Western Europe. Same-sex friends whose relationship was seen as contributing to societal welfare were buried in joint tombs in churches. But Bray finds that the relationships were always represented as nonsexual. Even in the case of the early nineteenth-century British heiress Anne Lister, whom we know from her diaries to have had sexual relationships with women, Bray argues that society, including her family and the local church, which blessed one of these relationships as a spiritual union, was not aware of its sexual nature. A few Hindu texts, however, I demonstrate, are able to endorse explicitly sexual same-sex relationships as virtuous and contributing to the greater good (see chapter 6).

Same-Sex Relationships and Hinduism Today

Since almost all the reported weddings in India over the last two decades, outside of the purview of gay movements, were conducted by Hindu rites, one question this book explores is: what elements in Hindu traditions enabled these marriages? To what extent have Hindu Gods, teachers, sacred texts, and sacred spaces been available to bless and sanctify same-sex relationships?

Many modern Hindu political leaders and some religious leaders take a negative view of same-sex relations, arguing, without evidence, that it is opposed to Hindu tradition and was unknown in premodern Hindu society. Their interpretation of Hinduism in this respect is no different from right-wing interpretations of Christianity and Islam. Hindu right-wing organizations like the Shiv Sena, RSS, VHP, and Bajrang Dal violently opposed the depiction of Hindu women in a lesbian relationship in the 1998 film *Fire*.

There is a gulf between these opinions and those of several modern Hindu spiritual leaders who draw on traditional concepts of the Self as unlimited by gender in their comments on same-sex relations. Sri Sri Ravi Shankar, founder of the international movement, Art of Living, states, "Every individual has both male and female in them. Sometimes one dominates, sometimes other, it is all fluid. There is nothing to feel bad about it." When asked about the high suicide rate among gay youth, tears came to his eyes and he responded, "Life is so precious. We need to educate everyone. Life is so much bigger. You are more than the body. You are the spirit. You are the untouched pure consciousness."[44]

Christopher Isherwood's guru, Prabhavananda, of the Ramakrishna Mission, was typical in viewing all desire as the same, and advising followers to see God in the beloved, thereby purifying love of lust. On hearing of Oscar Wilde's conviction, he remarked, "Poor man. All lust is the same." He advised Isherwood to see his lover "as the young Lord Krishna."[45]

This view is based on the idea that all beauty is a manifestation of divine energy and can lead the aspirant toward divine beauty. A similar idea is found in the Sufi concept of the human beloved as a reflector and witness of divine beauty.

When Ashok Row Kavi was studying at the Ramakrishna Mission in India, a monk discovered his homosexuality, and told him that the Mission was not a place to run away from himself, and that he should live boldly, disregarding negative opinions and testing his actions to see if he was hurting anyone.[46] Inspired by this advice, Row Kavi went on to become a pioneering gay activist, founding Humsafar Trust and gay magazine *Bombay Dost*.

On its website, the magazine *Hinduism Today* presents a mainstream view of Hindu attitudes to sex: "Intensely personal matters of sex as they affect the family or individual are not legislated, but left to the judgment of those involved, subject to community laws and customs. . . . Hinduism . . . does not exclude or draw harsh conclusions against any part of human nature . . . The only rigid rule is wisdom, guided by tradition and virtue."[47]

In a recent book, by Amara Dasa, a monk and founder of Gay and Lesbian Vaishnava Association (GALVA), several Gaudiya Vaishnava monks point out that since everyone passes through various forms, genders, and species in a series of lives, we should not judge each other by the material body but rather view everyone equally on a spiritual plane, and be compassionate to them the way God is.[48]

Those Indian LGBT and human rights activists who oppose the Hindu right's homophobia generally do not identify themselves as Hindu. This results in the media inaccurately depicting Hinduism itself as homophobic. This is beginning to change—when RSS chief K. Sudarshan made negative statements about homosexuality in 2004, gay activist Ashok Row Kavi wrote an open letter to him in the press, identifying himself as "a faithful Hindu" but "not a Sanghi [right-winger]," asking Sudarshan to read Indian history and ancient Hindu texts, and pointing out that not homosexuality but rather homophobia is a Western import.[49]

As I indicated earlier, some Hindu texts show the Gods actively enabling same-sex relationships. Others show same-sex marriages and marriage-like unions. For example, in the ancient epic, the *Mahabharata*, princess Sikhandini marries another woman. Later, Sikhandini is changed into a man. However, her marriage as a woman to a woman remains valid. When she becomes a man, she does not have another wedding ceremony. This is important, because it shows that marriage conducted by customary rites remains valid, regardless of gender. Modern writer Dan Detha emphasizes this in his version of a Sikhandin-type story (see chapter 3). Throughout this book, I examine how these premodern but still living traditions influence modern Hindu ideas of gender, love, and parenthood.

Historically, Hindu wedding ceremonies have never been uniform. There is no single Hindu hierarchy and no one leader equivalent to the Pope. As there is no absolute prohibition of same-sex sexual relationships in major Hindu texts, and same-sex marriage has not been a topic of extended debate, individual priests and spiritual teachers work out their own positions according to their reading, experience, and local practice (see chapters 4 and 10).[50]

Ayoni Sex and Kama: The Ambiguous Place of Desire

Ancient Hindu ascetic traditions, like ascetic traditions worldwide, tend to view all sexual acts with some distaste. This distaste arises in part from mistrust of physical pleasure as binding one to the phenomenal world, and in part from rules of purity and pollution. While procreative sex, hedged around with many rules, is enjoined on the householder, non-procreative sex is viewed with disfavor. As in most ascetic traditions, women are stereotyped as more lustful than men, and as temptresses of men.

These ideas influence householder life, which is structured as a set of obligations—to ancestors and to society. Many Hindu texts insist that everyone has a duty to marry and have children. If one renounces the world, one may be freed of this duty, but not otherwise.

However, these anxieties are countered in devotional practice and also in philosophy and literature, much of it composed by celibate teachers. These texts represent the Gods as erotic beings, and Kama (desire) as one of the four normative aims of life. In Hindu philosophy, nothing is absolutely good or evil, since everything is a manifestation of divine energy. The Gods are eulogized as present everywhere, in every plant, animal, and element, every part of the body, and every movement, thought, and feeling. It is in this spirit that the Shaiva (devotee of God Shiva) priest who married two women in Seattle in 2000 told me that he gives all the donations he gets for performing rites to the temple in his village in India, and thus offers whatever merit or demerit the rite may accrue, to Shiva, who absorbs it all.

Many Hindu Gods, like the Greek Gods, are embodiments of natural forces, like the sun, the moon, the wind, and are therefore relatively nonjudgmental figures. As the Bible remarks, the sun shines on the just and the unjust alike. Devotional traditions incorporate Goddesses as sexual beings and creative principles. Kama, God of love, is, in the earliest texts, a creative force inspiring desire and animation in the universe.

Ascetic, devotional, and philosophical traditions are intertwined; hence the often contradictory approach to sex within the same text. For example, the *Padma Purana*, a medieval text devoted to preserver God Vishnu both celebrates female erotic pleasure and elsewhere denounces it. In one somewhat comic story, when a husband consults his wife about how to hide treasure, she lectures him on the folly of asking a woman's advice. She describes in detail how rich widows masturbate, and have sex with anyone available. Her husband obediently asks her to go away, and hides the treasure by himself. The story then relates how the wife becomes so dear to Goddess Parvati that she is reluctant to return her to her husband.[51]

Ancient Hindu law books, which are permeated by purity and pollution concerns, declare *ayoni* or non-vaginal intercourse impure and punishable.[52] This category of *ayoni* sex is wide—it encompasses, among other things, oral sex, manual sex, anal sex, sex with animals, masturbation in the water or in a pot or other aperture. Yet the penalties prescribed are very light compared to penances, such as torture and death, imposed for other types of sexual misconduct, such as certain kinds of heterosexual adultery and rape. The *Manusmriti* exhorts a man who has sex with "a man or a woman in a cart pulled by a cow, or in water or by day" to "bathe with his clothes on" (11.174). The penalty for a man who has *ayoni* sex is a minor fine, the same as that prescribed for "stealing articles of little value" (11.165).

Modern commentators, such as Wendy Doniger, wrongly read the *Manusmriti*'s severe punishment for a woman's manual penetration of a virgin (8.369–370) as revelatory of that text's anti-lesbian bias. In fact, the punishment is exactly the same for either a man (8.367) or a woman who does this act, and is related not to the partners' genders but to the virgin's loss of virginity and hence of her marriageable status. The *Manusmriti* does not mention a woman penetrating a non-virgin woman. The *Arthashastra* prescribes a negligible fine for this act.

Ayoni (non-vaginal) sex by definition cannot include inter-vaginal sex, which is not listed as punishable in the law books. In chapter 6, I discuss some medieval texts' representation of inter-vaginal sex as a category that bypasses the prohibition of non-vaginal sex.

At first glance, *ayoni* sex as a catchall category appears comparable to the category of "sodomy" developed in medieval Europe and imported into India by the British. But the two categories are in fact very different. First, sodomy came to be constructed in Christendom as a horrific sin, almost the worst of all sins, "a favored synecdoche for sin itself."[53] Conversely, *ayoni* sex was a minor infraction of Hindu law. Second, sodomy came to be considered unspeakable, the sin not to be named among Christians, while no similar prohibition on mentioning *ayoni* sex developed among Hindus. Third, from the Renaissance until the nineteenth century in England and many other European countries, sodomy became not just a sin to be atoned for with religious penance but a legal crime to be punished with disenfranchisement, torture, and even death. No such development took place in the case of *ayoni* sex.

On matters of sex, including *ayoni* sex, Hindu law books appear to directly contradict other sacred texts such as epic and Puranic stories. The law tends to prohibit non-vaginal sex whereas the sacred stories often show heroic children and even deities springing from non-vaginal sex.[54] Often, the same text, for example, the *Mahabharata* and several of the *Puranas* (medieval compendia of stories about the Gods) contains both stories and precepts, and thus contradicts itself on the question of whether non-vaginal sex is impure or sacred.[55] The explanation of this apparent contradiction may lie in the fact that what is normally taboo or polluted may be sacred in special or ritual contexts.

In Hinduism, non-vaginal sex is not so much evil as forbidden or taboo. Like other taboos, it may be broken by divinities and those with divine powers or by ordinary people under special circumstances, with good results.

Kama and Eros

Like the ancient Greek God Eros, Kama, the Hindu God of love, is represented in poetry and paintings as a beautiful young man who shoots flower-tipped arrows that wound the heart. He rides a parrot or sparrow, emblems of sexual desire. Greco-Roman Goddess of love, Aphrodite/Venus also travels in a vehicle drawn by sparrows. Pigeons are associated with Kama and often appear in medieval miniature paintings of love scenes, including female-female love scenes. Similarly, turtledoves are associated with love in the West.

In the West today, Cupid is generally represented as a chubby, mischievous, near-naked baby, carrying bow and arrows. This image is a watered-down version of the ancient Greek God Eros, a beautiful, powerful young man. In the sixth-century BC, Sappho, a woman whom the ancient Greeks considered their greatest lyric poet, and who wrote many love poems to women, described Eros descending from heaven clad in purple, and striking her heart. There are several myths about Eros's parentage, one being that he is the son of Aphrodite. Despite the Christianizing of Europe, the living presence of these once-worshiped deities is found in words like "erotic" and "aphrodisiac."

Both Cupid and Kama underwent changes in the medieval period. Christianity gave a mystical meaning to erotic parts of the Bible, such as the Song of Songs, and sought to redirect amorous feelings from humans to God. But Eros and Aphrodite remained alive in the popular imagination, not just through Greek and Roman poetry, studied by every educated person and imitated by poets in modern European languages, but also through folk songs.

Kama was also transformed in medieval India along with the transformation of Hinduism by *bhakti* or devotion directed to a personal God. Medieval stories depict the conflict between ascetic celibacy and desire. The best-known story is that of Kama's battle with destroyer God Shiva.

After his wife Sati's death, Shiva becomes a celibate ascetic. The Gods send Kama to attract Shiva to Parvati, a reincarnation of Sati. Kama shoots an arrow at Shiva, which awakens him from his meditations. Shiva is so angry that he burns Kama up with fire from his third eye. This is generally interpreted as the defeat of eroticism by asceticism. But the story may also be read as Kama's victory, since Shiva, after he wakes up, falls in love with Parvati and marries her.

Kama, though burnt, remains an active but invisible force and acquires the name Ananga (the bodiless one). There are various stories of his resurrection. In one story, the Goddess revivifies him in a place called Kamarupa (Kama's form), in modern-day Assam in north-eastern India, where Sati's *yoni* (womb/vulva) had fallen. Here the Goddess is worshiped by the name Kamakhya (She whom Kama worships).

Same-Sex Union in the *Kamasutra*

The *Kamasutra*, a fourth-century sacred treatise on eroticism, categorizes humans into three groups—men, women, and those of the third nature. It further classifies men of the third nature into those who are masculine-appearing and desire other men, and those who are feminine-appearing and also desire men. It describes the lives and activities of men of the third nature (the occupations of hairdresser, masseur, and flower seller are recommended to them as ways of meeting men), and gives a detailed and sensuous account of oral sex between men. Several ancient texts, including Sanskrit plays, Hindu medical texts and Jain texts, develop a taxonomy of those who are inclined to same-sex desire (see Note on Methodology).

In a series of textual cruxes, the *Kamasutra* describes women's manly behavior during sex, which has been read by some translators as describing only female-male interaction, and by others as describing both female-male and female-female interaction.[56] In another important crux, the *Kamasutra* states that two men friends who have complete trust in each other may unite (II. 9. 36). The term used is *parasparaparigraham*. *Paraspara* means "mutual," and *parigraha* has many meanings, including "take in marriage," "have sexual intercourse," "take" "accept," or "seize."[57] Danielou translates it as "get married together," two Hindi translators as "have oral sex together," and Doniger as "do this service" for one another.[58]

I examined other uses of *parigraha* and its variants throughout the *Kamasutra*, and found that it is used eight times to refer to marriage; five times to mean "seize," "accept," "take" or "obtain" (as in seizing lips with lips, penis with vulva, or obtaining money); and six times to refer to copulation.

Apart from the reference to male-male union, there is only one other use of *parigraha* along with the word *paraspara* (mutual). *Parasparaparigrahayoha* (V. 4: 41) refers to a man establishing a mutual bond with another man's wife. *Paraspara* (mutual) is also used along with "enjoyment"—*parasparamranjayeyuhuhu* (V. 6: 1), referring to mutual sexual enjoyment between secluded women. Both these, like the reference to two men uniting, refer to unconventional types of union.

Most interesting are several uses of the word to refer to a woman who acts like a wife *parigraham cha charet* and *parigrahamkalpam* (VI.5: 36 and VI. 5: 4). These verses refer to a courtesan who behaves like a wife to a man who rewards her well. A courtesan's daughter whose hand is taken by a man acts as his wife for a year (VII. 1: 21–22), and the ceremony of taking her hand is termed *panigrahanavidhi*, which is also used in other texts to refer to marriage. *Parigraha*, unlike *vivaha* (which is fully sanctified marriage) can be used to refer to different types of marriage, including lower-status marriage, and also to lasting bonds outside traditional marriage, such as those between a man and another man's wife or those between a courtesan and her long-term lover.

So, to return to the verse that describes two men mutually uniting, the term *parigraha* there refers to mutual intercourse, but also carries the connotation of a union or bond of mutual acceptance, such as taking someone in lower-status marriage. It is important that this verse terms the two men *nagaraka*, that is, they are definitely not persons of the third sex but are men of the type constructed as normative in the text.

Modern Indians view the *Kamasutra* with a mixture of pride and embarrassment. Most Hindi translators insist that the *Kamasutra* catalogs varieties of sex only to warn the reader against them. One translator, Pandit Madhavacharya in 1911, blames homosexual men for the weakness of colonized India.[59] Another Hindi translator, Devadatta Shastri, also takes a homophobic view of same-sex desire, terming it perversion and a "bad act."[60]

It is a relief to return to the *Kamasutra*'s worldly, nonjudgmental tone: "It is by taking into account the country, the period, custom, the injunctions of the sacred texts, as well as one's own tastes, that one decides whether or not to practice these kinds of sexual relations. Practiced according to his fantasy and in secret, who can know who, when, how, and why he does it?" (II.9: 44–45).

I disagree with scholars who see the *Kamasutra* as an anomalous moment in an otherwise repressive history.[61] I demonstrate, especially in chapters 6 and 8, that liberatory ideas of sexual desire continued to circulate in later Indian texts, including versions and translations of the *Kamasutra* in other Indian languages.

Not By Sex Alone

The story of two girls from Kerala, Sree Nandu, 23, and Sheela, 21, made headlines across India in early 2004, after a tabloid accused them of lesbianism. The girls declare their love and commitment, saying they want to live together all their lives, but also insist, "Whether we are lesbians or not is our personal and very private issue. We did not invite anybody to peep into our private lives."[62] This is an important distinction—Sheela (who is from a Christian family) and Sree (who is from a Hindu family) describe themselves as "friends" and "lovers," but lifelong exclusive commitment or even marriage may or may not entail a sexual relationship.

Celibate marriage exists in both Christianity and Hinduism.[63] Several Christian saints foreswore sex with their spouses after conversion, and even today many married members of Hindu sects, such as the Brahmakumaris (God's maidens), renounce sex without always renouncing marriage. Some modern Indian spiritual teachers, including Shri Ramakrishna and Anandamayi Ma, lived in celibate marriages. As recently as the early twentieth century, several of Gandhi's followers entered celibate marriages, with his blessing. British writers Virginia and Leonard Woolf were happily married for decades, but gave up having sex almost immediately after their honeymoon.

In most societies today, a man and woman who marry are not legally mandated to have sex or to continue having it all their lives. Whether a husband and wife have sex, and what kind of sex they have is their private concern and is left to them. A same-sex couple should have the same right to privacy in marriage—whether they have sex, how often, and what kind, should be no one's business but their own.

Is Love for Heterosexuals Only?

The debate about same-sex marriage is also about who is entitled to use the language of love. Homophobic people claim that same-sex love is simply lust. Centrists, who would concede that it may be love, are still not sure whether it is as worthy of recognition as male-female love. At the other extreme, some gay activists and queer theorists argue that the language of love, like the language of marriage and coupledom, is too heterosexual, too commercialized, too patriarchal, too bourgeois (take your pick) for use by gay people. It is partly as a result of this type of argument that queer theorists in their writings overwhelmingly prefer words like "desire" to "love."

Lesbian and gay studies redirect public attention to the history of writing about love. Literary historians, from Oscar Wilde and Edward Carpenter in the nineteenth century onward, have repeatedly demonstrated that much of the mainstream language of love in the West, from Sappho and Plato through Michelangelo, Shakespeare and Byron to Dickinson and Auden, has indisputably been forged in same-sex fires, although later appropriated to cross-sex love alone.[64]

Women like Lalitha and Mallika or Leela and Urmila assume that the language of love is theirs to use. Among the items Lalitha and Mallika left behind was a greeting card showing a man and a woman kissing in silhouette against a sunset. This greeting card was not a new one; someone else had already used it. Mallika had pasted a piece of paper over the sender's name and written her own name on the paper. Inside was her message to Lalitha, giving her "a thousand kisses in public." Lalitha's note also stated, "The Rs. 25 placed in the diary is to be given as offering to Guruvayoorappan."[65] Guruvayoorappan refers to the icon of Shri Krishna in the temple at Guruvayoor, a famous temple town and pilgrimage site very close to the girls' hometown, Trichur.

Divine Blessing: For Heterosexuals Only?

The legend is that Guruvayoorappan's image was installed in the temple after it was miraculously rescued from drowning. The temple is a favored site for weddings, and, in the wedding season, dozens of weddings take place there every day. Lalitha's and Mallika's offering to this local deity perhaps contains within it both an allusion to their own marriage in death, and a plea to the deity with reference to their proposed suicide by drowning. The girls jumped off a ferry into the channel near Cochin, and were nearly drowned, but a crewmember and a fisherman rescued them.

Guruvayoorappan's temple is the site of heterosexual marriages, but is Krishna the God of heterosexual love and marriage alone? When two young women, Neeru and Meenu, married each other in the north Indian industrial town, Faridabad, on July 9, 1993, they did so in the local temple of Banke Bihari, with a priest officiating. Banke Bihari is Krishna in his pleasure-loving form, as lover of the cowherd girls in Vrindavan. Meenu, who is a singer at all-night religious worship sessions (*jagran*), and the sole supporter of her mother and three younger siblings, says of Neeru, "I am hers. I love her. This is a matter of the heart."[66]

Ideas of love and marriage are, for better or worse, inextricably connected with ideas of the sacred and divine. Is it, as some religious people claim, an entirely new idea and a sacrilegious one at that, for same-sex unions and marriages to claim divine blessing? Are Eros and Kama Gods only of love between man and woman? Does the Judeo–Christian God approve only of heterosexual love? At a less sacred level, is Saint Valentine's Day only for heterosexuals? These are questions largely outside the purview of the state, but tremendously important to many people.

In chapter 10, I briefly consider some living Gurus' and Hindu sects' views on same-sex love, and also some self-identified lesbian and gay weddings and unions, conducted by Hindu priests or with Hindu ceremonies.

Who Defines "Man" and "Woman"?

In 1996, the U.S. government passed a legislation called the Defense of Marriage Act (DOMA), which defines marriage as "a legal union between one man and one woman," as if it is self-evident who a "man" is and who a "woman" is. That it is far from self-evident is clear from the many cases of transsexuals both in India and in the West, who have legally changed their sex after gender reassignment surgery and then have married persons of their former sex. Such marriages have resulted in a number of knotty legal cases, some of which I examine in chapter 2.

Religious and philosophical traditions are generally more thoughtful than the makers of the DOMA when it comes to deciding what a man is and what a woman is. Both Christianity and Hinduism have taken a variety of approaches to the question of the spirit's gender and the question of divine gender. If God is ungendered or comprises all genders, if the spirit has no gender, or changes gender, or may have a gender different from that of the body, and if marriage is a spiritual, not just a physical union, the question of marriage becomes more complicated than the DOMA would have it.

Despite these complexities, opponents of same-sex unions in the United States constantly and triumphantly refer to God having created Adam and Eve male and female, as if that were the only thing God ever did. As several Jewish thinkers point out, if everything God created was good, then who created same-sex desire and those who experience it? Satan does not have the importance in Judaism that he does in Christianity, so ascribing such creation or perversion to him would not adequately answer this question for Jews, although it might for some Christians. Irshad Manji, a Muslim lesbian of South Asian origin, now living in Canada, asks why God created her a lesbian, and argues that Muslims should not judge one another; judgment should be left to God.[67] There are also many major cultures whose understanding of gender does not stem from an Adam and Eve model.

Performing Gender

Almost all the female couples who married each other in recent years presented themselves as bride and groom—in chapter 2, I discuss some of the legal and other implications of this self-presentation. In some cases, the groom was transsexual or transgendered; in others, she was merely a boyish or mannish woman.

Indian premodern texts about female homosexual relations also represent one of the women as assuming some type of masculine persona in the relationship (see chapters 6 and 8), while remaining a biological woman. This could be ascribed to the bias of the male authors of these texts. Conversely, if, as performance theory suggests, we all perform roles, including the socially constructed role of our assigned gender, and also perform a variety of roles in sexual relationships, these roles can be seen as performative rather than imitative.

Neither in the literary texts nor in the recently reported weddings does the masculine woman pass as a man. Rather, she performs a different type of gender identity as does the feminine woman coupled with her. Some nineteenth-century Indian texts stress this performativity, since the roles are described as arbitrarily assigned—two women break chicken wishbones or shell almonds; the one who happens to get a particular type of piece takes the masculine role and the other the feminine role (see chapter 8).

The converse—a masculine male and a feminine male forming a couple, is also represented in texts, such as movies (see chapter 9), and there is at least one recorded case of an Indian man passing successfully as a bride (see chapter 2).

Why More Women than Men?

Most of the couples who have married or committed suicide in India over the last two decades have been women. One reason may be that entering a cross-sex marriage and having anonymous same-sex encounters or liaisons on the side is relatively easier for a man than a woman. Most married men have greater mobility, leisure, freedom of social interaction, access to public spaces, and control over money, and less accountability to their spouses than women. They also carry less of the burden of housework and childcare than do women.

It is in the Indian urban upper middle class that one is most likely to find women as well as men leading double lives. In poorer families, women's lives are open to greater scrutiny than men's, and it is more difficult for them to carry on clandestine affairs. Living alone is more difficult for women from low-income families since hoodlums in slums see single women as easy prey. But if two women living together can be accepted as a family by the neighbors, they may be safer. In the professional middle classes, where people have more choices, more women than men choose to remain single. Women lose more freedom than men when they marry.

Many of the Indian women who elope together are from poor families, and their struggles call for amazing determination and courage. A recent example is that of Kajal, 24, and Nisha, 19, whose parents are construction laborers in Bhopal, central India. The two families live in a slum, and the girls were friends for years. Kajal was working as a peon in a school, and Nisha was unemployed. In April 2004, Nisha's parents took her to another state to marry her to a man. Kajal followed, and the two fled to Delhi. Nisha's family brought her back and forcibly married her to the man in May. Two weeks later, the two women disappeared again. When they reappeared in August, their families took them to the police, who advised them to see counselors. The women are reported to have told the police, "We will live together no matter what attempt is made to separate us," and did not turn up for their appointment with the counselors.[68] Had Nisha and Kajal been men or highly educated professionals, it is unlikely that the police would have intervened at all. As poor, uneducated women they have few resources or support structures outside the family.

Taking into account these gender inequalities, it still remains true that suicides and weddings reported in the papers are the tip of the iceberg, and not representative of all same-sex relationships or all homosexual and bisexual lives. I have anecdotal evidence of individual suicides by several middle-class gay men in India, shortly before or after their forced arranged marriages to women. A family can more easily cover up an individual's than a couple's suicide, since the latter proclaims a relationship and the former does not.

Gender Representation

There is an odd paradox of gender representation in Indian texts, which may or may not be related to the preponderance of women in reports of same-sex weddings and suicides. In premodern texts, male-male relationships far outnumber female-female relationships. But in modern India, this gets reversed—female-female relationships are represented in novels and short stories, by both men and women. Male-male sexual relations are rarely center stage; they tend to be placed in the underworld of prisons and slums, and dismissed as inconsequential. An important recent exception is the Hindi film *My Brother Nikhil*, 2005, which sympathetically depicts a committed gay male couple.

This modern focus on female relationships may be in part because, from the nineteenth-century onward, the Indian media has been influenced by social reform and nationalist movements to take a sympathetic view of women as victims, whose struggles are justified by their oppression. Almost all the media reports on female-female weddings from the 1980s onward project the women as victims of oppression (for example, Leela was represented as an unfortunate widow and Urmila as a victim of child marriage to an incompatible husband), strongly suggesting that their lesbian marriages are a protest against injustice rather than an expression of love for each other. Since there is no long-standing discourse in India on the oppression of homosexuals (no Indian Oscar Wilde dramatized homosexual suffering for the Indian public), it is harder for the media to project homosexual men as victims.

Second, Indian nationalists picked up from colonial rulers the view that India's weakness was the result of Indian men's deficient masculinity; in conjunction with modern construction of male homosexuals as effeminate, this led to male-male relations being seen as not just an individual and familial but a national disgrace.

Rural to Urban: Same-Sex to Cross-Sex Bonding

Historians have shown that urban cultures have been more hospitable to same-sex relationships as life-choices, and to the development of networks among those who make these choices. Evidence from ancient Rome, medieval Europe, and medieval China, Japan, and India demonstrates this. Modern gay culture also flourishes in cities, where greater anonymity and freedom from family pressure is available to people.

Modern Indian cinema represents the shift from rural to urban. Male-male friendship in the village stands for "authentic" Indian culture, which is disappearing, and male-female love in the city for modernity.[69] This transition is clearly seen in *Yarana* (Friendship), 1981, where the hero, a single man, is devoted to his male friend. The friend takes him to the city, where his rustic wit confounds everyone's attempts to do a Pygmalion on him. But in the second half of the film, the heroine, a city girl who always wears Western dress, declares her love to him, and angrily remarks that his preference for same-sex friendship over love shows that he is an uneducated fool.

Hearing this, he instantaneously gets transformed into an Elvis look-alike, singing a song that proclaims, "The whole era is intoxicated by women . . ."[70] Thus the cultural transition to modernity is linked to the necessity for a shift from romantic friendship to romantic heterosexual coupledom.

Birth versus Choice: Indian Concepts of Predilection

One of the most contentious and also most futile debates around gay rights in the West centers on the question: Is same-sex desire chosen or predetermined at birth, biological or socially constructed? Antigay forces insist that it is chosen, and that gay people should choose to become heterosexual. In response, many gay people now insist they were born gay, and have no choice in the matter. Given how many people experience both cross-sex and same-sex desire, often switching halfway through life, sexologist Alfred Kinsey's argument that most people are bisexual, with some inclining more to same-sex, some more to cross-sex, and some equally to both types of desire, seems more plausible.[71]

The fact that no choice is entirely free does not mean that the concept of choice is meaningless. The way genetic and social determination may work with rather than against the concept of choice is suggested in some Indian concepts of love. Indian narratives tend to represent love not as accidental but as an expression of perfect suitability. In Indo-Islamic narrative, choice in love is not shaped by individual idiosyncrasy but is predestined. In Hindu narrative, choice is shaped by patterns of action, thought, and feeling developed in this life as well as earlier lives. One term for these patterns is *samskaras*. The closest English-language equivalent would be the notion of conditioning. But since *samskaras* are the result not just of one but of several lives, they encompass both conditioned and innate (what we might call genetic) tendencies.

Another, related meaning of *samskara* is a rite of passage, a rite through which a person attains completeness. Marriage is termed a *samskara*. Through marriage, a person is supposed to fulfill and express the individual self, which is itself a product of *samskaras* or accumulated habits of attachment.

The paradox of conditioned choice is encapsulated in the premodern institution of the *swayamvara* (literally self-chosen) ceremony, in which Hindu women from princely classes choose their grooms. In one form of the *swayamvara*, the woman, accompanied by her girlfriends, walks around the hall where prospective bridegrooms are assembled, and garlands the one she chooses as her husband. Here, her choice is circumscribed by class and caste status as well as by her father's choice of whom to invite.

But another type of *swayamvara* circumscribes her choice even further, because her father arranges an archery contest, and she must garland the winner. Why is this garlanding still read as a choice? Because the cultural assumption is that she will love the man destined for her, who is therefore bound to win the contest. It is for this reason that the two ancient epics represent heroines Sita and Draupadi falling in love at first sight with the man who wins the contest, and "choosing" him in the *swayamvara*. Some women choose their grooms before the formal ceremony. In an ancient and still very popular legend, princess Damayanti falls in love with King Nala after hearing him praised by others. When he comes to her as a messenger from the Gods, who also want to marry her, she tells him plainly that she will choose only him in the *swayamvara*, and will marry no one but him.

The idea that an individual, even a woman, is entitled or even obligated to exercise some type of choice, more or less circumscribed, in sexual partnership, and that this choice has validity, because it represents the tendencies of a self that has passed through many births, may be invoked in situations otherwise deemed illicit, such as adulterous relations, cross-caste and cross-class relations, and same-sex relations (see chapter 4).

Many of the great medieval Indian love stories are about Muslim lovers—Sohni-Mahiwal, Heer-Ranjha, Sasi-Pannu. These narratives present the lovers as predestined for each other—those who oppose their love oppose not just an individual choice but a force of destiny far stronger than social forces. In most such stories, the lovers are separated and destroyed by society, but survive in collective memory.

Occasionally, the lovers may successfully wed, as in the legend of twelfth-century Hindu king, Prithviraj Chauhan (a historical figure) who loves Sanyogita, daughter of his rival, King Jai Chandra. Jai Chandra does not invite Prithviraj to the *swayamvara*, and as a further insult, has a statue of him dressed as a watchman, and placed at the door. Sanyogita walks past the invited princes and garlands the statue. Her defiance would be symbolic except that Prithviraj has disguised himself as a watchman and taken the statue's place. As soon as the garland is placed round his neck, the statue comes to life and carries Sanyogita away. The happy ending is only temporary because Sanyogita's father remains hostile and colludes with the Afghan invader, Muhammad Ghori, to defeat Prithviraj in battle.

Recently, two women enacted something similar. The newspaper report, titled "Woman abducts her lady love," called it "an 'abduction' story with a difference."[72] In Yeotmal, Maharashtra, Suraj Shukla dressed as a man and took a job as a security guard at a local bank to "lure away" Swati.

The two eloped to Nagpur to get married. The police picked them up and, on interrogation, Shukla confessed to being a woman.

Chosen and Given Families

Parents pressuring children to marry often warn them that, without a family, they will have no one to look after them in old age. This assumes that procreation is the only way to build a family. In fact, however, traditional families are simultaneously given and chosen—they incorporate and construct kin in many non-procreative ways. They also incorporate choice—few people are on equally loving terms with all members of their given families. We all choose to be closer to some relatives than others; and most families experience temporary or permanent rifts between members. So when gay people today build chosen families with partners, ex-partners, friends, and their own or their friends' adopted or biological children, this is new but not entirely new. I therefore use the term "chosen" rather than "alternative"—gay families and friendships are not alternatives to a reified model of given biological family; rather they bring to the surface the complex process of choice and biology that constitutes any family.

Traditional Indian families may incorporate not just grandparents, widowed aunts and uncles or orphaned cousins but also family friends and elderly servants, on whom kinship terms are bestowed. Chosen kinship (termed fictive kinship by some anthropologists) permeates Indian culture, and is highly elastic, ranging from the casual to the highly serious. Younger people address family friends, neighbors, and even strangers by terms like aunt, uncle, father, mother, grandfather/mother, and now the Indian English "Auntyji/Uncleji." This was the practice in the West until about fifty years ago—witness the American painter universally known as "Grandma Moses" (1860–1961).

In north India, people frequently create sibling ties with cross-sex friends through the annual *rakhi* ritual, wherein women tie auspicious threads on brothers' wrists. Same-sex friends are also incorporated into the family by terming them siblings. Parents often term a child's same-sex partner their son or daughter. In most Indian languages, cousins are termed siblings, leading to the terms "cousin brother/cousin sister" in Indian English. In Tamil, paternal aunts and uncles are addressed as "big" or "small" father and mother. Similarly, children often address their father's other wives in Hindi as "big mother" or "small mother."[73]

This type of incorporation has limits, though—the fictive child or sibling may inherit responsibilities and some rights, but will rarely inherit property. Yet it can be very adaptable—man-woman relationships that begin as chosen sibling (*rakhi*) relations sometimes develop into romance and marriage. It also spills out of domestic into public spaces. For example, when a gay man in Delhi was unexpectedly hospitalized, his chosen family members, about a dozen people, gay and straight, claimed to be his cousins and were allowed into his room. This happened even though the hospital is an elite one, catering mainly to Western tourists.

Humsafar Trust, a gay organization in Bombay, annually celebrates the festival of Rakhi with the tying of threads on friends' wrists, regardless of gender, and also Bhishma Parva (a day named for the childless celibate great-uncle in the epic *Mahabharata*), with commemoration of gay people who have died that year, including victims of AIDS. This is an example of the way gay people build chosen kinship networks and rituals to sanctify them.

This type of kinship building can be traced back at least to the nineteenth century in the West. Oscar Wilde's inner circle, which supported him after his disgrace, comprised straight and gay people, single, coupled and married people, friends, lovers, and ex-lovers, both male and female. Katherine Harris and Edith Bradley ("Michael Field"), mentioned earlier in this chapter, counted among their closest friends the long-term male couple Charles Ricketts and Charles Shannon, who ran a press that printed the works of Wilde, Field, and other Aestheticist writers. Terry Castle has examined the friendship between lesbian writer Radclyffe Hall and gay dramatist Noel Coward, and Martha Vicinus the friendship networks of gay people in the nineteenth century.[74]

Some thinkers, including pioneering poet-philosopher Edward Carpenter (1844–1929), in his book *The Intermediate Sex* (1908), argue that gay people, precisely because they less often have children, contribute more to society, putting their energies into social welfare disproportionate to their numbers. Saint Paul advised early Christians not to marry so that they could devote themselves to God rather than to their spouses. Gandhi encouraged his followers to treat the entire community as their families and all children as their children. It is perhaps an updated version of this paradigm that some activists today appeal to, when they urge gay people not to get co-opted into families but rather to develop new ways of relating to others.

Even Saint Paul and Gandhi, however, convinced very few to give up the pleasures of family for those of communal welfare. Nor are the two necessarily incompatible. It is possible to care for given family and chosen family, and still retain nonfamilial concerns.

Same-sex weddings of the Leela–Urmila type can be integrated into families precisely because of this continuum. Marriage is a relationship with which everyone is familiar, and extending it to same-sex couples is not so different from extending familial relationships to non-biological friends.

Toward Flourishing

Whether or not governments recognize same-sex unions, these unions exist where they seem least likely—in the bosom of heterosexual marriage. Thousands of apparently heterosexual, apparently happily married men and women in every country maintain simultaneous long-term unions with persons of their own sex. Sometimes, these arrangements are embedded in conventional families, when co-wives or sisters-in-law maintain long-term unions, with or without the collaboration of husbands. In chapters 6, 7, and 8, I discuss both premodern and contemporary representations of such unions.

Same-sex unions intertwine with cross-sex marriage in other ways, when homophobia and familial pressure drive gay people to drastic measures. Many Indian lesbians and gay men today are choosing to wed one another, in "marriages of convenience," in order to satisfy their parents, while simultaneously maintaining their same-sex unions. In chapter 7, I discuss personal advertisements placed by such individuals in South Asian gay media and on websites and chat lists.

The most drastic measure is the decision to commit suicide together, hoping to unite in a life to come. In chapter 4, I examine how the idea of rebirth (a Hindu idea that pervades Indian social consciousness) affects attitudes to socially disapproved unions. Rebirth works as a metaphor for indestructibility. In the *Inferno*, Dante meets condemned lovers, heterosexual and homosexual. Even in hell, they retain both their love and their dignity. He meets other heterosexual and homosexual lovers in Purgatory, where they repent their desires. If, however, the spirit is thought of as born not once but many times on earth, its attachments persist from one life to the next. Transforming or eradicating these attachments is not a matter for social engineering. Hence many Hindu teachers advise their followers to work through and transcend their desires rather than suppressing them. In this view, suppression is counterproductive because the attachment will be reborn, even stronger. Self-regulated expression, on the other hand, conduces to the type of life, which, in another context, Aristotle termed *eudaimonia* (flourishing).

Chapter 2

Who Decides?: Marriage Law, the State, and Mutual Consent

The church does not in fact, marry anyone. People marry each other.

—*Episcopalian Bishop John Shelby Spong, 1990*[1]

I loved her and I couldn't live without her, so I decided to change my sex and marry her.

—*Manish/Manju Chawla, 1989*[2]

On March 21, 1993, two women, Vinoda Adkewar, 18, and Rekha Chaudhary, 21, from neighboring villages, went to the Registrar of Marriages in the town of Chandrapur, Maharashtra, in western India, and declared their intention to marry. The Registrar said he was "perplexed and unable to decide what to do next"; he told them to return after some days. Judicial officers and police held "urgent deliberations," and obtained legal opinions. On April 13, while a crowd waited outside, they spent three hours dissuading the women from "such an unusual alliance." Finally, Vinoda agreed to return to her parents. Rekha, enraged, threw into a ditch the red bridal sari she had brought for Vinoda, and walked away with tears in her eyes.[3]

Much like the city officials across the United States who were thrown into consternation in spring 2004 by same-sex couples demanding their legal right to marry, these Indian officials did not immediately turn the women away. This hesitation results from ambiguities in the law, which make it hard to refuse two individuals the right to marry.

Is Same-Sex Marriage Legally Possible?

When two nurses, Jaya and Tanuja, married in 2001, a government advocate commented: "Indian laws have no provision of marriages between the same sexes [*sic*]."[4] To ask if there are special laws for same-sex marriage is to frame the question incorrectly. Rather the question is this: is it possible for same-sex couples to marry under existing marriage laws? My reply is that it is. The law does not unambiguously exclude same-sex couples.

The U.S. federal government and many state governments would not have felt the need to pass Defense of Marriage Acts (and constitutional amendments in some states), defining marriage as between a man and a woman, if they were sure that the law already excludes same-sex couples. It is precisely because same-sex marriage is conceivable within the framework of current marriage laws and customs that governments feel the need to change those laws.

Until fairly recently, Hindu laws were construed to exclude intercaste marriage and remarriage of widows and divorcees, while Christian law in England, the United States and Canada prohibited divorce as also interreligious and interracial marriages.[5] Today, most Hindus and Christians would consider these interpretations incorrect, and would agree that such unions can be accommodated within sacramental marriage.

Marriage is not one thing for cross-sex couples and another for same-sex couples. The basic features that characterize marriage do not by their nature exclude same-sex couples. Among these are the partners' mutual consent, and community consensus recognizing the pair as married.

So, when a same-sex union is solemnized with customary rituals, is it a marriage or not? It may not be a marriage in the eyes of the government but, as the Rabbi who performed the Jewish part of my wedding ceremony said, the spouses may be wed "in the eyes of God and all enlightened people."

Mutual Consent—Essential for Marriage

The Urdu saying, *Miya Biwi razi, kya karega qazi*? (When husband and wife agree, who needs a judge?) indicates that marriage is a mutual agreement between two individuals. In Muslim marriage, the officiant (who may be any adult Muslim male) asks bride and groom if they consent to the marriage. When they say they do, the marriage is complete. In the Hebrew Bible, when Jewish patriarch Isaac seeks his first cousin Rebecca's hand in marriage, her brothers ask for her consent. When she consents, they bless her and send her off to Isaac. The two live together, and are considered married (Genesis 24).

A Christian historian of marriage writes, "What is actually essential for marriage . . . is a very simple form of mutual consent."[6] In Christian marriage, as well as many forms of civil marriage, the basic formula consists of the question, "Do you take so-and-so to be your husband/wife?" followed by an affirmative answer. Consent is crucial; everything else is a nonessential ceremony.

Ancient Hindu law recognizes two to twelve types of marriage, some more socially approved than others, but all valid. Eight types came to be recognized in later texts.[7] Among these types, the *gandharva* is marriage by mutual consent.[8] *Gandharva* marriage requires no parental consent, no rituals, no officiant, and no witnesses. The ancient legend of Shakuntala represents perhaps the most famous *gandharva* marriage in Indian literature. Shakuntala marries a king without witnesses and he later repudiates her and her son. Despite the risks involved, the fourth-century *Kamasutra* depicts a nurse telling a young woman the story of Shakuntala, in order to persuade her to elope with her lover (III. 5. 5.) While some law books disapprove of *gandharva* marriage, much ancient and medieval Indian literature terms it superior to other forms.[9] The *Kamasutra* (III.V.29–30) states that *gandharva* is the best form of marriage because it is based on attraction (*anuraga*).

In some ancient texts, lovers marry with no witness except fire, which is a manifestation of God. In the third to fourth century Sanskrit play, *Avimarakam*, the hero secretly enters the heroine's bedchamber, walks around a fire with her, and then declares that since they have taken seven steps with fire as witness, they are now married.[10] Modern Hindi films depict this type of union—when pregnancy results from one-time premarital intimacy, lovers wed in secret by exchanging garlands and praying to God (see *Aradhana* [Worship], 1969).

One article on the 1987 marriage of policewomen Leela and Urmila describes their wedding as a *gandharva vivaha*.[11] Most Indians would readily understand this as the equivalent of the modern "love marriage."

Apart from the spouses,' is anyone else's consent required? Ancient Hindu law books consider the best marriage to be one arranged by two families, where rites are performed, and the father gives his daughter as a gift to the bridegroom, along with other gifts. This is similar to the Christian custom in which the bride's father walks her to the altar and "gives her away" to the groom. In Muslim marriage, too, the groom and the bride's father register their agreement, and the groom gives or agrees to give *mehr* (dower) to the bride's father.

Other commentators, however, insist that even in the Vedic (ancient Hindu) ceremony, the bride is a gift from the Gods, not from her father, and the wedding is valid by mutual consent of bride and groom. Parental consent is not essential to the validity of a Hindu, Muslim, or Christian marriage.

Community versus State

Communities sometimes recognize marriages that governments refuse to recognize. Even in Western democracies, where obtaining a license from the government is essential for a legal marriage, community consensus continues to be important. This is clear in the categories of common-law marriage and domestic partnership that have evolved in many Western democracies that give some of the rights and responsibilities of marriage to a man and woman who have lived together for some time, and whom the community recognizes as married, even though no ceremony has taken place and no marriage has been registered. In England, this type of marriage predates marriage registered with church or state. It is also found among some tribal communities in India, where a man and woman are considered married if they set up house together.

Both in India and the West, many communities treat same-sex couples as if they are married. Partners live together and share a financial and social life; they entertain together and are invited out together. In 1999, an Indian magazine carried a story about two women, Santosh, 32, and Manju, 33, living in Patel Nagar, Delhi (see photo 2.1). They met as nursing students in 1984, fell in love, and started living together "as man and wife, and the people of Patel Nagar have taken it in their stride. . . . the residents have accepted the 15-year-old 'marriage.' "[12] There was, however, no wedding ceremony: " 'We did not marry in the conventional sense, it was more of an emotional one where we accepted each other as life partners,' said Santosh." It is from the existence of such couples, seen by many people as married although no ceremony has taken place, that the concept of same-sex domestic partnership evolved in the West.

The Indian government recognizes as legal any marriage performed according to customary rites, whether or not a license has been obtained. This is a crucial difference between Indian and Western marriage law. The vast majority of Indians never inform the state when they get married or divorced. The state documents their marriages in other contexts, such as taxes, property registration, children's education, bank loans, and passports, but relatively few couples obtain marriage licenses.

A typical example of the way marriage, divorce, and adoption occur outside of state purview is that of S, a poor Brahman woman who works as a domestic

Photo 2.1 Santosh and Manju at home, Patel Nagar, New Delhi, 1999. Photo: *The Week.*

servant in Delhi. She was engaged as a teenager, and when her fiancé died, she was married to his older brother who already had a wife. Since she did not have children for some years, she adopted her co-wife's daughter. When she later had her own children, she still retained financial responsibility for this girl. After the husband's death, the two women lived in adjoining apartments inherited from him. S got her adopted daughter married, but the girl returned to her natal family within a few weeks. S and community elders negotiated with the groom's family; the dowry was returned and S got the girl married to another man. That marriage too was conflicted, and the girl now spends about half her time living with her biological mother, and the other half in her marital home.

Even in the West, customary religious marriage may be used to pressure the state into granting legal recognition. This happened in a path-breaking case in Toronto. Under the Ontario Marriage Act, any couple may be granted a marriage license if a church, following ancient tradition, reads the marriage banns for three consecutive Sundays prior to the wedding. In 2001, the Metropolitan Community Church in Toronto read the banns and married two men, Kevin Bourassa and Joe Varnell, and two women, the Vautours. The couples filed a case, asking the state to register their marriages. The Court of Appeal ordered the province to register the marriages. Ontario began issuing marriage licenses to same-sex couples in June 2003, and the government of Canada promised to follow suit.[13] This government recognition of a church ceremony as sufficient for marriage is similar to the Indian government's recognition of a customary religious ceremony as sufficient to constitute a legal marriage.

State Recognition of Marriage: How Essential?

Many people wrongly think that the state has "always" regulated marriage. In fact, the state's takeover of marriage is relatively recent. In premodern Europe, the Church considered marriage a secular matter so priests did not perform weddings; they simply blessed them as they did other secular undertakings, such as sowing and harvesting of crops or opening of new workshops. In the thirteenth century, Pope Innocent III decreed that both spouses' free consent is the sole essence of marriage. In church law, a verbal contract in the present tense between a male of 14 or older and a female of 12 or older, witnessed by two persons, was a marriage.[14] A priest might bless the married couple, usually at the church door, but this was not essential. People could marry each other anywhere with God as witness.

A similar idea is found in Hindu weddings, where individuals marry each other with the fire as witness. The officiant (who is often a priest but need not be; anyone who can read the verses correctly will do) invokes various Gods as witnesses, and leads the spouses and their families through the rituals and vows.

Under pressure from influential families in medieval Europe, the blessing at the church door gradually became a blessing at the altar inside the church. This developed into the couple taking the sacrament together. In the middle ages, the Church increasingly began to take control of people's sexual lives. Thus began the gradual move to the priest performing a wedding ceremony, and recording it in a church register. This is the origin of the modern practice of the state registering marriages.

Royal and aristocratic weddings were arranged as political alliances, but the state had little to do with ordinary people's weddings. When Henry VIII broke away from the Roman Church, he agreed with the Pope that Church and state were inseparable; hence he declared himself head not just of the British state but of the Church of England. From the Reformation onwards, governments began to increasingly assert control over marital matters but the Church still retained primary control.

The state's gradual takeover of marriage was completed only after the French Revolution, when the Republic in 1792 decided that the only valid marriages would be civil ceremonies performed and registered by a government officer. The religious ceremony, if any, is irrelevant to the legality of marriage, and has to be performed after, not before, the civil ceremony.

Variations in Democratic Marriage Laws

Several democracies followed France's example—in Belgium, Germany, Hungary, Italy, the Netherlands, Rumania, and Switzerland, the civil ceremony alone is recognized by the law, and in most of these countries clergy are prohibited from performing the religious ceremony before the civil marriage has taken place.

However, some countries took a different route. England and Wales, under the Marriage Act, 1949, recognize not only civil marriage but also a religious marriage performed by banns in the Anglican Church, which is the established Church of England. This recognition of religious marriage is similar to Indian law's recognition of religious marriages, except that Anglican weddings are recorded in church registers and Hindu marriages are usually not recorded in writing (Indian Muslim marriages are recorded in a *nikahnama*).

In Denmark, Norway, and Sweden, ministers of recognized religious denominations must register with state authorities and also obtain a certificate from the government for each marriage before they perform it—the religious ceremony is then also a civil ceremony. In the United States, in addition to state recognition of the ministers, each couple must obtain a marriage license before the marriage. In England and Wales, for any religious wedding ceremony (apart from one conducted by the Anglican Church) to also be a civil ceremony, not the officiant but the premises (churches, mosques, temples), must be registered with the state. Couples are also required to follow the procedures for a civil marriage.

A wedding that follows religious law but breaks the State law is illegal even if performed by a registered officiant or on registered premises. For example, U.K. law prohibits bigamy and most Muslims in the United Kingdom disapprove of it, but some Muslim bigamous marriages have taken place in mosques in the United Kingdom. These weddings are illegal. In India, Muslim law is recognized by the state, so a Muslim bigamous marriage is legal, but since the state changed Hindu law in 1955, Hindu bigamous marriages after 1955 are illegal.

Indian marriage law is distinct from Western democracies' laws, insofar as in India a religious marriage is legal even if neither the officiant nor the premises is registered with the state, and the couple has not obtained a license. This means that virtually any Hindu priest in any temple can legally marry two people.

State Takeover: How Democratic?

Many consider the state's take-over of marriage progressive because it supposedly returned marriage to the secular arena by eliminating Church control. Some would argue that it is more democratic for the state to regulate marriage than for communities, who may be more or less patriarchal, to do so. What is often forgotten is that state takeover severely restricted individuals' long-standing right to marry by mutual consent. Marriage by mutual consent, a truly ancient form of marriage, has been supplanted in the West by a two-centuries' old law, giving absolute control to the state, which is also a patriarchal institution. The Hawaii Supreme Court refused to recognize common law marriage on the ground that it infringes on "the state's role as the exclusive progenitor of the marriage partnership."[15]

Indian democracy, although in most respects similar to Western democracy, provides an alternative and, in my view, somewhat less authoritarian model. Indian marriage law tries to maintain an uneasy balance between a central state-defined law and regional community laws. When the British colonized India, they found a bewildering array of marriage practices, in contrast with the legal uniformity by then established in England. The British attempted to codify Indian law; they passed different marriage and family laws for Hindus, Muslims, and Christians, which are sometimes termed "personal laws."

The independent Indian government continued this practice, but state governments were allowed to modify the application of these laws. This is similar to the situation in the United States, where family laws fall within the purview of state governments, not the federal government. Thus, some states in the United States allow first-cousin marriage and others do not, but all states recognize marriages performed in other states (until 2004, when several states began refusing to recognize same-sex marriages performed in Massachusetts).

Under the Hindu Marriage Act, the state recognizes any Hindu marriage as legal that is "solemnized in accordance with the customary rites and ceremonies of either party thereto."[16] Couples do not need to obtain a license. Muslim marriage is a contract, not a sacrament. Indian Muslim marriages do not have to be registered with the state nor does a license have to be obtained. Individuals of any religion can contract a civil marriage under the Special Marriage Act, 1954, for which they do need a license.

Not Going to the Chapel: Quaker and Jewish Marriages

Prior to 1753, the British government did not recognize any marriage not performed by the Church of England. Therefore, Quaker and Jewish marriages were not recognized as marriages. This is because the state recognized only one religion as legitimate—that of the Anglican Church. Quakers and Jews did marry and considered themselves married. Were these marriages or not? They were married, in the eyes of "enlightened people," but not in the eyes of the state. Looking back today, we would say that these unions were marriages, even though the state did not recognize them.

A similar question arose in India when the nineteenth-century Hindu reformist sect, the Brahmo Samaj, began performing a simplified Hindu wedding ceremony. In 1868 a court declared these marriages invalid. To remedy the situation, the first civil marriage law in India, the Special Marriage Act, was passed in 1872. It created a huge controversy; those arguing in its favor pointed out that since so many forms of Hindu marriage already existed, the Act was just adding another one.[17]

The U.S. federal government's current refusal to recognize same-sex marriages performed by ministers and rabbis or by civil officials in Massachusetts, and the Indian government's refusal to recognize same-sex marriages performed by Hindu priests, places these marriages in a situation analogous to that of Quaker and Jewish marriages in the eighteenth century or Brahmo marriages between 1868 and 1872.

The U.S. federal government, because it is secular, cannot explicitly base its refusal on a particular interpretation of Christianity, but government officials justify their refusal by referring to God as well as Judeo–Christian tradition. It is clear that their interpretation of God and Judeo–Christian tradition differs from that of many churches, ministers, and rabbis who perform same-sex weddings.

A Hindu Shaiva priest I spoke to in 2002 said he knew that other priests in his lineage would be shocked by his officiating at the marriage of two women. Having thought about it, however, he became convinced it was the right thing to do, because marriage is a union of spirits, and Hindu texts clearly state that the spirit is neither male nor female.

Community, Custom, and Law

All law originates in custom. Muslim and Christian laws are to some extent fixed by written texts that claim to conform to a holy book. Even so, there are major variations, for example, Muslim Sunnis argue that Mutaa or temporary marriage was valid at one time but is no longer so; Muslim Shias argue that it is still valid. Indian law recognizes this difference of opinion, and upholds the validity of Mutaa marriage among Shias.

The role of custom is most clearly apparent today in Hindu law. All ancient Hindu law books state that custom is powerful and overrides written texts. Although different schools of Hindu family law, such as the Mitakshara and the Dayabhaga, are based on different interpretations of sacred texts, commentators repeatedly state that custom and approved usage override written texts.[18]

The British attempt to make Indian law uniform, by dividing it on the basis of religion, so that Hindus were governed by Hindu law, Muslims by Muslim law, and Christians by Christian law, erased the eclectic mix that actually existed in the practices of many communities. For example, the Khojas of Maharashtra converted about 400 years ago to Islam yet continued to follow Hindu rules in matters of inheritance. But in 1937, under British rule, the Muslim Personal Law (Shariat) Application Act was passed, bringing all Muslims, including the Khojas, under its purview.[19]

But even the British were compelled to recognize the importance of custom in Hindu law because Hinduism has no one holy book that overrules other holy books. In 1868 the Privy Council ruled: "under the Hindu system of law, clear proof of usage will outweigh the written text of the law."[20]

In the 1950s, the Indian government continued the process of making laws uniform by passing laws to regulate and reform Hindu marriage, divorce, and other family matters. There was widespread opposition to this codification, primarily on the grounds that it would erase the diversity of customary practices that had the force of law among different Hindu communities. In deference to this diversity, the government built limited recognition of custom into family law.

Many Indians today consider this recognition of custom backward and divisive. There is a strong movement afoot today, supported by many political parties and women's organizations, both on the right and the left, to pass a uniform civil code for all Indians, erasing differences in marriage practices between Hindus and non-Hindus, as well as among Hindus.[21]

In my view, Indian law's recognition of custom, although it generates problems, is nevertheless valuable because it retains a balance between centralized state and localized community. It ensures that the state does not have the exclusive right to define marriage.

How Old Must a Custom Be?

Custom, being fluid and constantly changing, is notoriously difficult to define. Indian courts have sometimes defined custom as prevailing from "time immemorial," but have modified this unrealistic criterion to "long-standing" practice. The Indian government recognizes as valid any marriage performed according to community custom. Whether or not a Muslim or Christian wedding conforms to custom is relatively easier to decide than whether a Hindu wedding does so, because Hindu custom, even within the same local community, is far from uniform. Some Hindu weddings are customarily conducted over five days, others over three days, most over several hours, and some in a few minutes. Among the Vaishnas in Bengal in the nineteenth century, exchange of garlands was the only wedding rite.[22]

From the nineteenth century onwards, some Hindu communities began trying to render their customs uniform. Powerful communities such as Agarwals, Kayasthas and Lingayats formed nationwide bodies, whose leaders met and decided what their customs were. Upwardly mobile communities gradually changed their customs to conform to upper class and caste practice.

New groups, such as the Arya Samaj, also defined themselves as communities and formulated new practices that they defined as legal "customs." The anti-Brahman movement in Tamil Nadu, south India, instituted a new type of marriage known as self-respect marriage. In 1967 the Tamil Nadu government passed a law recognizing as valid any marriage performed by the groom tying a *tali* (wedding pendant) on the bride in the presence of witnesses.[23]

But community members do not always change their customs to conform to leaders' views. When questioned in court, community members give widely divergent accounts of custom. Courts tend to recognize as valid a custom that is mentioned in several cases over a period of time. If a group that defines itself as a community performs a ritual repeatedly over a period of time, it may convince the courts that this ritual has become a custom.

At least one Hindu teacher, Pandit Shailendra Shri Sheshnarayan Ji Vaidyaka, has argued that gay people could be seen as constituting a separate community (see chapter 10). If several same-sex weddings take place in a Hindu community or, alternatively, if gay people who are Hindus conduct several same-sex weddings over a period of time in a particular region, same-sex marriage could come to be legally defined as customary there.

Can "Bride" and "Groom" Be of the Same Sex?

The Hindu Marriage Act states, "A marriage may be solemnized between any two Hindus, if the following conditions are fulfilled."[24] The list of conditions prohibits bigamy, insanity, marriage before the age of 21 for the groom and 18 for the bride, and certain forms of biological relationship between the two, unless these forms are permitted by community custom.

The gender of the "two Hindus" is not stated. However, gender is assumed and appears in the third requirement, that "the bridegroom has completed the age of twenty-one and the bride the age of eighteen at the time of the marriage." The terms "bride" and "groom" appear many times thereafter in the Act. When Vinoda and Rekha, "two Hindus," sought to marry, Vinoda was 18 and Rekha was 21. The two women explicitly stated that Vinoda was the bride and Rekha the groom.

In most of the lesbian weddings reported in India over the last two decades, one woman presented herself as the groom and the other presented herself as the bride. Several couples performed the rite of the groom putting vermilion (*sindoor*) in the bride's hair parting. Some female grooms undergo or say they intend to undergo a sex-change operation. Others have no such intention but have short hair, and wear shirts and trousers. In the few photographs available of joint suicides, though, both women are in female dress.

When two women in India publicly claim the right to marry (as opposed to privately marrying each other in death) they seem to rest this claim in part on their presentation of themselves as a couple in which one woman is the bride and the other the groom, even though both are female. The degree to which family and community accept this claim appears to be inseparable from the degree to which they accept the marriage.

In the case of Neeru and Meenu who married in 1993, Neeru uses the male alias, Dinesh Sharma, and dresses like a man. However, she is biologically female. Her family married her to a man, but she left him in a few days. In the case of the two policewomen, Urmila always dressed in shirt and trousers, and had short hair, while Leela wore a sari and jewelry when off-duty, and had long hair. In photographs, Urmila looks like a boyish girl rather than a man. Her family refers to her as a female and by her female name, yet they treat Leela like a daughter-in-law, saying they have a responsibility to look after her. Some communities are thus able to integrate female-female marriage into their interpretation of Hindu law, by recognizing one woman as the groom and the other as the bride.

This does not always work, however. Raju, who married childhood friend Mala in December 2004, has short hair, wears jeans and leather jackets, and has a male-sounding name while Mala wears red bangles, a symbol of marriage. After their marriage by a Hindu priest in Delhi, they returned to their hometown, Amritsar, where Raju told reporters, "We have vowed to live together for the rest of our lives as husband and wife." Mala threatened to commit suicide if they were forcibly separated, and said, "I have left my family for her."[25] But their families and neighbors remained extremely hostile and boycotted them, so they had to go into hiding.

Can a Woman Be a Groom?

When I married a woman in June 2000 in New York, my partner jokingly told her mother, "You'll again be the mother of the bride" (since my partner's sister was already married). Her 87-year-old mother (a Jewish-German-American) replied, in surprise, "*You're* the bride?" Implicit in this question is the assumption that a wedding, even a wedding of two women, involves a bride and a groom, not two brides. The idea that gender makes both women brides while the requirements of marriage make them bride and groom appears in the caption of one story about the Leela-Urmila wedding: "Bride Grooms Bride."[26] The more masculine-appearing (butch) partner is perceived as the groom, and the more feminine-appearing one as the bride. This perception is relative—it is conceivable that the butch one would be seen as the bride if she were with an even more butch woman.

Indians, like Shakespeare's audiences, are familiar with cross-dressing. They know that a man can play a bride and a woman a groom. In Hindu religious drama, enacted in many parts of India during festive seasons, this assumption of a persona is not seen as mere playacting. Performance of religious drama confers sacred status on the actors, who are seen as temporarily embodying the divinities. Audiences often worship the male actors who play the role of Rama and his bride Sita.

If two persons experience their performance of bride and groom deeply and seriously as the truth of their life, must observers accept their view? Can a woman be a husband throughout her life? As discussed in chapter 4, one Hindu reading of same-sex relationships is based on the assumption that the partners were cross-sex lovers in a former life. In that case, the groom, though biologically female, may be perceived as male in spirit.

An interesting twist on the question of "bride" and "groom" arose in San Francisco in 2004 when marriage licenses were issued to 4037 same-sex couples. Unlike officials in Oregon who had same-sex partners fill out the regular form, San Francisco officials printed a new form, which replaced the words "bride" and "groom" with "Applicant 1" and "Applicant 2." Among the many legal challenges to these marriages was the claim that they were not marriages because the words "bride" and "groom" were removed.

Male Brides

In one of two reported cases of a Muslim male-male marriage in India, Mustafa, 22, was the bride and Harfan, 28, the groom. They married in February 2004 in Garhmukteshwar, Ghaziabad, north India. The report does not describe the ceremony or say whether the bride passed as a woman, but says that friends and relatives "thought the marriage was a joke."[27] Harfan was already married to a woman. It appears that he was making ingenious use of his legal right as a Muslim male to remarry, but when the couple brought home a flower-decked bed of the kind customarily used for wedding nights, friends and relatives stripped and beat them up. The report ends, "this controversy does not seem to have deterred the couple that plans to live together as man and wife."

Another case of Muslim male-male marriage is recorded in a documentary film entitled *Terhi Lakeer* (Crooked Line, 2002), made by Aparna Sanyal, Amrit and Arunima, students at the Mass Communications Research Center, Jamia Milia Islamia University, Delhi. This film refers to the marriage of Naseem, a Muslim man, with Vijay, a Hindu man, by Hindu as well as Muslim rites. Naseem dressed as the bride and the person officiating at the *nikah* thought he was a woman. Later, Naseem succumbed to family pressure and married a woman, and the two men broke up.

In 1999, a Pakistan newspaper reported that an organization called *Tanzeem-e-Murtian* in south Punjab had facilitated many male-male marriages since 1988, performed as *nikah* by a Maulvi, popularly known as Maulvi Disco, who was married to a woman and had three children, but was also gay: "After *nikah*, one male partner is declared as husband and the other as wife in a gay gathering."[28] In 1998, the Maulvi admitted that the organization covered rural areas and small towns in the area from Faisalabad to Rawalpindi. But, after "ruthless state persecution," the organization went underground and the Maulvi denied conducting male-male marriages. A police case was filed against the Maulvi and ten other members of the organization, but the court acquitted them for lack of sufficient evidence.

Socially versus Legally Recognized Marriages

Unlike Vinoda and Rekha, most Hindu female couples do not try to get their marriages registered with the civil authorities. Instead, they get married by a ceremony of a type prevalent in their communities. Their marriages are socially recognized. If challenged in court, could a case be made for their legal validity?

Since Hindu priests, unlike Muslim and Christian officiants, generally do not document weddings in writing, if a Hindu wedding's validity is challenged in court, judges examine evidence like photographs, and testimony of priests, witnesses, and guests. The general principle seems to be a common sense one: if it looks like a wedding, if it is seen and understood to be a wedding, it is a wedding.

The question of whether a particular customary ceremony does or does not constitute a valid marriage has arisen in cases of Hindu men marrying second wives. In 1955 the government changed Hindu law to enforce monogamy for Hindu men, and made bigamy a criminal offence, punishable with imprisonment and a fine. Women who would have been second wives under earlier Hindu law were now mistresses with no legal status or rights. Their children, who would have been legitimate in Hindu law, became illegitimate. By the 1950s monogamy had already become the norm for most Indians. However, some Hindu men continue to remarry without divorcing their first wives. Either the first wife or the second wife may then sue the man for bigamy, although often neither does.[29]

When the first wife sues, she paradoxically tries to prove that the second wedding was a valid marriage because it was conducted according to community custom, and is therefore punishable. Exactly the same paradox arose in the United States in 2001, when a Utah court sentenced a Mormon, Tom Green, to five years' imprisonment for bigamy. Green married and divorced five women in turn, but continued to live with all of them. Divorce and remarriage are not crimes, nor is it illegal for unmarried adults to cohabit, but the state illogically argued that all five were his common-law "wives," so he had committed bigamy.

A Hindu man accused of bigamy generally tries to prove that the second ceremony was incomplete and was therefore not a wedding but some other type of union. Thus, in the precedent-setting case of Bhaurao versus State of Maharashtra, 1965, Bhaurao married Indumati in 1956, and then married Kamlabai in 1962. He was convicted of bigamy, but when he appealed to the Supreme Court of India, his conviction was overturned. The Supreme Court was convinced by Bhaurao's argument that the second marriage omitted two essential Hindu ceremonies, invocation before the fire and *saptapadi*, and was therefore not valid.

The prosecution argued that the second marriage was a *gandharva* marriage. The ceremony included a *puja* and exchange of garlands. Witnesses from Bhaurao's community testified that they had attended other weddings solemnized by these ceremonies. The judges, however, declared, "The marriage between Appellant No. 1 [Bhaurao] and Kamlabai does not come within the expression 'solemnized marriage.' "[30]

By referring to the Bhaurao-Kamlabai union as a "marriage" that was not a "solemnized marriage," the judges inadvertently acknowledged the existence of two types of marriage—one that is both legally and socially recognized, the other that is socially but not legally recognized.

Higher and Lower Status Marriages

Before the British codified marriage laws, Indian communities ranked marriages, using different terms to distinguish higher from lower status marriages. This system provided some recognition to lower-status unions, such as second marriages, which now have no legal validity.

Among Shia Muslims, *mutaa* marriages generally have a lower status than *nikah*. Among Hindus, remarriages of widows and divorced women had a lower status. Ancient texts refer to a remarried woman as a *punarbhu*. Although a wife (*vadhu*) for practical and inheritance purposes, she did not have the ritual and social status of a first wife (*grihapatni*).[31] Even in the United States, remarriage after divorce often has a lower social status than a first-time marriage.

Several celebrities in India have high-profile bigamous marriages, which are not legal, but are socially recognized. Classical dancer Raja Reddy married his wife Radha's sister, and lives with both women and their children. Film stars Dharmendra and Hema Malini married and had a child, even though he already had a wife.

Civil unions and domestic partnerships in the West, hailed by some gay activists as "radical alternatives" to marriage, are in fact nothing but lower-status marriages, in the sense that they confer some but not all of the state-bestowed benefits that civil marriage does. If they conferred all the same benefits, they would merely be marriages by another name.

Socially, both in the West and in India, even when families and communities recognize same-sex unions as marriages, residual heterosexism ensures that they rarely accord a same-sex couple the same status as a cross-sex couple. The difference in status appears in subtle ways, such as the amount of financial assistance parents give, the type of gifts given, and the way partners are introduced.

Same-Sex Marriage Not Illegal

Can the democratic state prevent people from entering into same-sex unions or punish them for doing so? Unlike bigamy, same-sex marriages are not punishable in India or the West. Even in societies (such as the United States before 2003 or India today), where certain same-sex sexual acts are illegal, same-sex marriage is not illegal because marriage is not equivalent to the performance of any particular sex act.[32]

Even the police seem to recognize this distinction. Harfan's relatives handed him and his spouse Mustafa to the police, but the police refused to arrest them, because while sodomy is a crime in Indian law, same-sex marriage is not. Similarly, when two Nepalese women textile workers, Sita Malla, 24, and Rupa Shrestha, 16, got married in August 1998, they were arrested but later released because "although same-sex marriage may offend some social expectations, it's not actually illegal."[33]

The lesbian unions reported in the Indian press have generally been termed "marriages" by the women themselves and by the journalists reporting them. Some journalists put the term in quotes to indicate the ambiguity: " 'Wedded' women cops to challenge sack."[34] One journalist refers to Leela and Urmila as "legally wedded";[35] another extremely homophobic article opens with the statement, "Three months ago two women were legally married to each other in Jodhpur, followed by two others a month later, where families of both sides even blessed the couple."[36]

Friendship Agreements

People all over the world are devising creative ways to give their unions, both cross-sex and same-sex, legal status short of marriage. In Europe, these have evolved into a variety of civil unions and domestic partnership arrangements, granted some degree of state recognition. In Gujarat, western India, in the 1980s, some businessmen began to draw up "*maitri karar*," or friendship agreements to confer financial rights on women who would have been their second wives under old Hindu law, but were mistresses under new Hindu law.

Some Indian same-sex couples also enter into such contracts to endow each other with status and rights. In 1987, two women teachers, Aruna Sombhai Jaisinghbhai Gohil, aged 31, and Sudha Amarsinh Mohansinh Ratanwadia, aged 29, got a "friendship contract" registered in Baroda, Gujarat. They stated that they had known each other since 1978, when they were at a teachers' training school. They had been living together all these years and were now both based in Vadadhali village. Since they intended to continue living together, they decided to enter into a friendship contract. The newspaper report stated that this was the second such contract between women registered in Baroda district court.[37] In Orissa in 1998, Mamata and Monalisa entered into a similar contract (see chapter 4).

These procedures are comparable to the practice in the United States of same-sex couples drawing up wills and powers of attorney to confer rights on one another. Friendship agreements evolved independently in India under Indian contract law, which recognizes any contract, whether notarized or not, between consenting adults, if it does not violate state policy. The idea of a friendship agreement is also based on premodern traditions of recognizing friendship as an institution (see chapter 5).

Legal Evidence of Marriage?

Same-sex couples also use various other means to acquire legal validity for their marriages. In 1998, Shweta and Simmi, both aged 22, signed an affidavit and got it notarized at the civil court in Patna, Bihar, in eastern India. The affidavit stated that they had married at the temple of Lord Mahavir on April 4, "in the presence of witnesses according to Hindu rites." It continued, "They have been living and enjoying conjugal life together as husband and wife and will maintain their relationship till their death." They attached photographs and copies of documents that proved they were adults.[38]

As distinct from a friendship contract, this affidavit, preceded by a traditional wedding, represents the two women as husband and wife. The court notary agreed to witness the document, as it stated proven facts. Like a civil union, this union is modeled on the idea of marriage as contract rather than sacrament, and is thus closer in some ways to the Jewish or Muslim concept of marriage than to the Hindu or Christian concept. The combination of this contract with a sacramental Hindu marriage constitutes a creative alternative to registration of civil marriage.

Same-Sex Weddings and Customary Ceremonies

On April 27, 2001, two women, Jaya Varma, 25, and Tanuja Chouhan, 32, got married in a Hindu ceremony at Mahamaya temple in Ambikapur, Bihar. "The couple took the traditional vows as a priest chanted the *mantras*. They went seven times round the sacred fire to solemnize their marriage."[39]

At the same ceremony, Jaya's sister was also married, to a man. Although Jaya's sister's marriage is validated by the Indian state and Jaya's is not, Jaya's family and community and the Hindu priest validated both equally. About a hundred people were present at the reception. Jaya's entire family was present.

A week after the ceremony, the couple went to get the marriage registered. Maninder Kaur Dwivedi, the registrar, listened to their arguments, but refused to register the marriage. The president of the Bar Council said, "The Hindu Marriage Act of 1955 will not recognise this as a marriage. This is illegal. However, this is not a crime." Arvind Singh, government advocate, commented: "One of the main requirements of Hindu marriage is the necessity to procreate, have children. The question of which does not arise here."[40]

Is Procreation Necessary?

Arvind Singh's claim that the ability to procreate is necessary for a valid Hindu marriage is parallel to U.S. right-wing leaders' claim that same-sex marriages are invalid because procreation is the purpose of marriage. Some U.S. courts have reiterated this specious claim, overlooking the fact that the law in most countries, including India and the United States, allows people who are well over the age of procreation to get married. The law also allows women who have had hysterectomies or men who have had vasectomies to marry.

Under the Hindu Marriage Act, a spouse cannot ask for divorce or nullification of a marriage on the grounds that the other spouse is sterile. However, s/he can ask for annulment if the partner cannot sexually consummate the marriage because of impotence. The court will hear the petition only if the petitioner did not know about the impotence at the time of marriage. In Swaraj Kumar Grover versus Sudershan Grover, Justice T.P.S. Chawla of Delhi High Court noted, "Hindu Marriage Act 1955 does not recognize sterility of the wife as a ground for divorce."[41]

If an Impotent Man Can Marry, Why Not a Transsexual?

Indian law on sex change is confused and unclear. *Hijras* often have great trouble in declaring themselves female when obtaining government documents like passports. At least one person in India has tried to legally challenge a marriage on the grounds that the union could not result in procreation.

In 1987, Tarulata, 33, underwent a sex-change operation and became a man, taking the name Tarunkumar. He then married Lila Chavda, 23, by a civil and a religious ceremony in December 1989. Lila's father, Muljibhai Chavda, a schoolteacher, went to the Gujarat High Court, asking for the marriage to be annulled on the grounds that Tarunkumar can neither have "natural" sexual intercourse nor procreate.

Muljibhai's lawyer argued, "Even an impotent Hindu male can marry because impotency is no bar to his marriage. In this case Tarunkumar was not a Hindu male at the time of his birth."[42] He also invoked the anti-sodomy law, Section 377 of the Penal Code, claiming that this was a lesbian relationship. Lila's father had no legal standing to ask for the marriage to be nullified, since under the Hindu Marriage Act only the spouses or guardian of a minor spouse can present such a petition. Yet, the court issued notices to the Registrar of Marriages and to the doctor who performed the sex-change operation, asking them why the petition should not be admitted.

Tarunkumar and Lila claimed that Lila's father was upset because in his community the groom's family pays a dowry to the bride's family (the reverse of the mainstream practice where the bride's family pays a dowry to the groom's family), and Tarunkumar had not paid him a dowry. The couple was quoted as saying, "Even if the court declares our marriage null and void, we shall continue to live together because we are emotionally attached to each other."[43] This statement emphasizes the way same-sex unions exist in the interstices of the law—neither recognized nor criminalized.

Transsexual Marriages: Do Chromosomes Equal Gender?

Because of the definitional ambiguity of the categories "man" and "woman," governments that attempt to limit marriage to unions between a man and a woman always fail to outlaw at least some same-sex marriages. Most Western democracies now allow people to undergo gender reassignment surgery and legally change their sex. Some conservative courts, refusing to recognize these gender identity changes, inadvertently end up validating same-sex marriages. Thus, when a Texas court decided in 1999: "Gender is fixed by our Creator at birth,"[44] the unintended consequence was that a man who becomes a woman by sex-change surgery and lives as a woman, can legally marry another woman in Texas by producing a birth certificate that states she is male. In September 2000, a lesbian couple, Jessica Wicks and Robin Manhart, were issued a marriage license in Texas because Robin, now a woman, was male at birth. Oddly enough, this means that two people who look as if they are of the same sex can marry in Texas, but two people who look like a man and woman, cannot marry, if one of them is transsexual.

It also means that some states do not recognize marriages that are valid in other states. Christie Lee completed gender reassignment surgery (male to female) in 1979, and her driver's license and birth certificate were altered to reflect her new sex. In 1989 she married Jonathan Littleton in Kentucky, and they moved to Texas. Jonathan died after six years, and Christie brought a wrongful death suit against his doctor. When the lawyers discovered her past, they moved to void her marriage. In 2000, the Texas Supreme Court decided that Christie's marriage, valid in Kentucky, is invalid in Texas, because "Male chromosomes do not change with either hormonal treatment or sex reassignment surgery."[45]

On the other hand, some states, such as California and Florida, recognize that gender reassignment surgery can change a person's sex. Because California has passed a Defense of Marriage Act, forbidding same-sex marriage, a woman who has become a man cannot marry another man in California, but can marry a woman, as the partners' legal sex would be different, although their chromosomal sex would be the same. By Texas' definition, this marriage, allowed in California, is a same-sex marriage.

Kansas courts were divided on the issue till the Kansas Supreme Court, like the Texas Supreme Court, decided in favor of gender determination by chromosomes.[46] In 2002, the Kansas Supreme Court declared the 1998 marriage of J'Noel Gardiner, a finance professor, to Marshall Gardiner, a former stockbroker, invalid because J'Noel, who had gender reassignment surgery in 1994, "was born a male and remains a male for purposes of marriage under Kansas law."[47]

The U.S. Supreme Court has not yet ruled on these issues but, however it defines gender, it cannot avoid the quandary of legalizing some type of "same"-sex marriage. If only chromosomes count, then a biologically male person can legally and socially become a woman and marry another woman. If gender reassignment overrules chromosomes, then a biologically male person who has legally become a woman can marry a man. Furthermore, if a person changes their sex while married to a person of the opposite sex, it would be very difficult for the state to invalidate their marriage, even though it has become a same-sex marriage. This is because the general rule is that a marriage can be declared null and void only if one partner demands dissolution or dies. The only logical way out of these quandaries is for the state to recognize marriages, regardless of the partners' gender.

The U.S. immigration service has recently announced that it will not recognize any transsexual's marriage (whether to a man or to a woman) for purposes of immigration!

Hijra and Other Male-Male Marriages in India

Hijras are female-identified males, some of whom are transgendered, others transsexual, and a few intersexed. They take female names, and dress in female garments, but their body language, gestures, and occupation clearly identify them in public spaces as *hijras*, not women. Many are homosexually inclined men from low-income families. While many *hijras* get ceremonially castrated after joining the community, not all do.

Hijras live in their own groups, hierarchically organized as families and headed by a senior hijra who is a mother or guru figure, and earn a living by dancing and singing at lifecycle celebrations such as weddings and births. They bless babies and newly wed couples, and are paid for their services because their curses are feared. Some of them also beg and engage in prostitution. Some men live with or spend considerable periods of time in *hijra* communities, without themselves becoming *hijras*.

Often, *hijras* couple up with men and live as their wives. A play titled *Hijra* shows a man marrying a *hijra*'s adopted son, who passes as a woman in order to emigrate as his wife.[48] Occasionally, *hijras* also pair off with one another, and/or adopt children (see chapters 6 and 9).

In 2002, one *hijra* tried to get her marriage to a man legally recognized. Durga Ghosh, a 30-year-old *hijra*, married Gourab Roy, an 18-year-old boy, at a Kali temple in Orissa. The Registrar of Marriages did not object to the sex of the bride but to the age of the groom. He refused to register the marriage because the boy was under 21, the minimum age of marriage for men. Under the Hindu Marriage Act, if the spouses continue to live together and neither of them nullifies the marriage, it becomes valid when the groom turns 21.

Durga remarked that Gourab's parents were unwilling to accept the relationship, so she had tried to commit suicide: "However, I was rescued by Gourab who then decided to marry me."[49] Gourab worked as a courier boy and was afraid of losing his job due to the publicity. Durga said she would support him, and offered him all her assets, including a house and money.

Family Support: The Crucial Element

Hijras generally leave their natal families and live in a community, which functions as a chosen family, headed by a senior *hijra* who is treated as mother and teacher. But most young women who resist marriage to a man and want to live with a woman are in danger of losing family support, with no alternative community to turn to. In almost all the reported cases of female-female weddings, the couple was lower middle class, and had the support of at least one of their two families. Family support is not necessary when the partners are educated and employed, and living in big cities, although even in such cases, family support is desired, and, when forthcoming, helps integrate the couple into the community.

In India, the family, not the state, is still the primary source of support for most people. The state does not give welfare and other benefits to the unemployed, although it does provide free health care and education of uneven quality. Relatives, biological and marital, have to support orphans, widows, the unemployed, disabled or depressed persons, and the old and infirm. This support may be given grudgingly or generously, but it is rarely entirely withheld.

Young people, especially women, find it difficult to strike out on their own without family support, because few other financial and social support systems are available. With family support, one can often defy many social and even state prohibitions. Daughters of wealthy families, whose parents, especially fathers, encourage them, can and do lead highly unconventional lives, both personal and professional.

In middle and lower-middle class families, socially forbidden marriages successfully take place when families choose to support the couples, but run into immense trouble when families oppose them. In one case in the 1980s, a young Hindu woman in Delhi, who was a volunteer at *Manushi*, the women's organization of which I was a founding member, fell in love with a Muslim man. Although her family was allied with the rightwing Hindu organization, the RSS, she managed to win over most of them, even an uncle who was an RSS activist, and they all participated in the wedding. This was possible because of the already high status of women in this family, and the good relationship she had with her parents.

If the family of even one partner is supportive, they can help the couple combat the hostility of the other partner's family. This was clear in the case of Madhu and Manju. Manju Chawla, 22, underwent a sex-change operation in January 1989, and became a man, Manish. On February 8, 1989, Manish married friend and classmate Madhu.[50] Manish explained that they had been in love for some time, and that he had had the sex change surgery in order to marry Madhu. Manju/Manish's family was ambivalent about her boyish behavior when she was young but later accepted it, and decided to support her sex change.

Madhu's father, a wireless operator, remained unremittingly hostile to his daughter's marriage and complained to the police that Manish had abducted Madhu. The police arrested Madhu and produced her in court, where her father argued that she was too confused to take any decision, so the woman judge sent her to a Nari Niketan, a state-run institution for homeless women. Manish's father took up cudgels on the couple's behalf. He got a stay order from the High Court, and Manju was again produced in court where she testified that she wanted to live with her husband, so she was released. The couple then moved in with Manish's family. On their own, the couple would likely not have had the wherewithal to resist Madhu's father.

In the cities, some women who obtain education and employment do manage to live in couples, without family support. Santosh and Aruna, discussed earlier, were both nurses. Santosh, who is known as *bhai* (brother) in the neighborhood, earned well as a contractor, small-time politician, and local heavyweight. They now run a business together and have little contact with their natal families.

For poorer women, living alone or even as a couple is not always safe. Working women's dormitories and other types of housing for employed women provide safe environments for many female couples. But most women, and even many men, who resist cross-sex marriage, continue to live with their parents. This is true even of many highly educated and very independent people in big cities. This can be a comfortable arrangement for both parents and children. Daughters who marry are expected to move to their husbands' homes, but sons who marry are, sooner or later, expected to live with or take care of their elderly parents. Despite this convention, especially prevalent in north India, married daughters do often take care of their parents. Thus, when female couples move in with the parents of one woman, they follow an accepted pattern of parent-child relations.

It is when natal families and communities turn actively hostile to young people's same-sex relationships that suicides may occur, a pattern I examine in Chapter Four. In the West, dependence on natal families is much less crucial, especially for adults.

The difference between the Western and the Indian situation is not absolute but relative—in India, the natal family can often enlist governmental and police help to coerce young adults against their will. While some individuals fight back and establish their independence, many others, as I discuss in later chapters, succumb and either submit to heterosexual marriage or commit suicide.

Chapter 3

Is the Spirit Gendered?: Fluid Gender, Sex Change, and Same-Sex Marriage

"The Wise One, who sees the same everywhere, sees no difference between happiness and sorrow, man and woman, fortune and misfortune."

—Ashtavakra Gita, *XVII: 15*

"After all, what is marriage? It is a wedding of two souls. Where in the scriptures is it said that it has to be between a man and a woman?"

—Sushila Bhawasar, village school teacher, commenting on the marriage of her neighbors, policewomen Leela and Urmila[1]

Sushila's view of marriage as a union of two souls would be accepted by most Hindus in India and also by many Christians in the West. However, not all would agree with her conclusion that since the soul is not gendered, a marriage between two men or two women is permissible. In this chapter, I discuss the implications of the soul's genderlessness for the possibility of same-sex marriage, and examine some traditional ideas of human-divine same-sex marriage. While these concepts refer to levels of reality beyond day-to-day embodiment, I argue that they are available to people who respond to the present-day phenomenon of same-sex marriage.

In Hinduism, as in Christianity and Islam, gender is often perceived simultaneously as very powerful and as irrelevant. This paradox makes possible the enforcement of gendered social roles along with the perception of spiritual non-difference. Thus, although St. Paul declares that in Christ there is "neither male nor female" (Galatians 3:28), he assigns different roles to women and men. How does this paradoxical understanding of gender affect an understanding of same-sex unions on the spiritual and social planes?

Original Wholeness

In the *Symposium*, Plato's fifth century BC dialog on love, Aristophanes recounts a myth about the origins of gender and sexual desire. He says that originally all human beings were of three sexes—male, female, and male-female or hermaphroditic. Each of these original beings was round. It had two faces, one in front, one at the back, and two sets of genitals. It also had four legs and four arms. Aristophanes describes these original beings not as monstrosities but as stronger and more complete than present-day humans: "Terrible was their might and strength, and the thoughts of their hearts were great."[2] When they challenged the Gods to combat, Zeus decided to reduce their strength by cutting each one in half. Thus originated human beings with one face, one set of genitals, two arms, and two legs.

Erotic love, Aristophanes says, is our search for our original other halves. Those men who were originally part of a round, all-male being, desire other men. Those women who were originally part of a round, all-female being, desire other women. But those who were originally parts of round, hermaphroditic (male-female) beings desire the opposite sex. Aristophanes thus explains heterosexual and homosexual desire as originating in the same way—from human desire for an original state of wholeness. Same-sex desire is linked to being all-male or all-female, and cross-sex desire to androgyny. This is the opposite of modern stereotypes of homosexual men as effeminate, lesbians as masculine, heterosexual men as manly and heterosexual women as feminine.

Western commentators have generally viewed this remarkable narrative as anomalous, and even dismissed it as farce because comic dramatist Aristophanes narrates it. However, comparing it to stories in ancient Hindu texts reveals similarities that suggest a possible common Indo-European source for some elements of the narrative—original roundness and power; the creator's reduction of this power by splitting the round beings; original androgyny splitting into maleness and femaleness.

In the *Kurma Purana*, creator God Brahma produces Rudra, who is "half-male and half-female and was too terrible to behold. 'Divide yourself,' saying this Brahma vanished out of fear" (I. 11. 3).[3] Rudra splits into a male and a female. The male is Shiva and the female Parvati, who is then born as the daughter of the king of mountains. She longs to unite with her original other half, Shiva. At her father's request, Parvati reveals herself in her divine form: "It had hands and feet all round, had eyes, heads and faces in all directions" (I. 12. 59).[4] On seeing this form, her father is "frightened and struck with awe" (I. 12. 200).[5] He requests her to reveal another form, and she assumes a gentle female form with two eyes and arms. In this account as in Plato's narrative, original roundness and multiple limbs inspire terror in father Gods.

Gender Resulting from a Split

In Plato's myth, one of the three types of original round beings is male-female. This type of being is absent from the creation story of Adam and Eve, accepted by Jews, Christians, and Muslims. In Hinduism there are many creation stories, and some incorporate ideas of androgyny and bisexuality. Although Hindu Gods and Goddesses are male and female respectively, Hindus also think of every deity as simultaneously male and female, or neither male nor female. Thus, in the *Kurma Purana*, Goddess Parvati tells her father that she is "non-different" from Shiva (I. 12. 91).[6] All the texts devoted to Goddesses stress this non-difference.

Shiva is often represented as half-male and half-female (*ardhanarishwara*). The common understanding of this icon is that Parvati is one with him and constitutes half of him. Here, as in Plato's myth, heterosexual union is understood as congruent with androgyny. Both Plato's myth and the Hindu texts associate heterosexual union with androgyny or the condition of being half-male, half-female.

Goddess texts show this androgyny originating from original all-female and all-male forms. Thus the *Lalita Mahatmya*, a Goddess text embedded in the *Brahmanda Purana*, subordinates Shiva to an original female principle. It tells us that the male Shiva attained his androgynous form through devotion to the Goddess: "By worshipping and propitiating her and also by means of the power of meditation and Yogic practice, Lord Siva became the leader of all Siddhas and also became the lord, half of whose body has the female Sakti form" (5: 30).[7]

Many-Limbed Beings

Plato's narrative is one of very few texts in the Western canon that imagines an entity with more than two arms and two legs as a positive figure. Hindu Gods, Goddesses, and other divine beings frequently have more limbs than humans—Brahma has four heads, Gayatri has five, and other deities are routinely represented with four or more arms. This multiplicity of limbs is a sign of divine strength and versatility. In the *Kurma Purana*, the universe itself is many-limbed: "All round it [the universe] has hands and feet; it has eyes, heads and mouths on all sides; all round, it has ears; it exists enveloping the world" (II. 3. 2).[8]

Apart from residual images like that of the six-winged holy beings "full of eyes before and behind" in the Book of Revelation (4:6–8), Christian icons represent divine beings anthropocentrically as two-armed and two-legged. Yet, Christianity retains the older idea of circularity as perfection. The Christian wedding ring and the Hindu wedding garland inscribe this notion into marriage. Dante imagines Paradise as a series of circles, and British poet Henry Vaughan (1622–1695), drawing on Ptolemaic cosmology, envisions eternity as circular: "I saw eternity the other night/Like a great ring of pure and endless light,/All calm as it was bright . . . "[9]

Love and Perfection

The idea that love leads to spiritual perfection is found in different forms in mainstream Christianity, Christian neo-Platonism, Islamic Sufism, and devotional Hinduism. There are different ways of iconographically representing the perfection made possible by love. The circle is one such symbol; another is the embrace that fuses. In Plato's *Symposium*, Aristophanes says that lovers experience bliss when fused in sexual intercourse because this temporarily returns them to their original state of wholeness (round beings with four arms and four legs). He also claims that lovers long for everlasting fusion: "the intense yearning which each of them has toward the other does not appear to be the desire of intercourse, but of something else which the soul desires and can not tell. . . . And the reason is that human nature was originally one and we were a whole, and the desire and pursuit of the whole is called love."[10]

Some Hindu icons visually represent this perfect fusion. The *ardhanarishwara* form of Shiva and his wife Parvati shows them fused by love into a being that is simultaneously male and female. A same-sex variation of this is Harihara, an icon that is half-Shiva and half-Vishnu. The two male Gods are fused by love into one being, which has attributes both of destroyer God Shiva and preserver God Vishnu.[11] These icons signal the ultimate unity of the divine, and the irrelevance of gender (male and female are one; male and male are one; preservation and destruction are one). But the icons also signal love. Parvati is famous for her devoted love of Shiva.

In another love story, Shiva asks Vishnu to assume his female form as the enchantress Mohini. When Vishnu does so, Shiva falls in love with the transformed Vishnu/Mohini and embraces him/her. The offspring of their union is the God Ayyappa or Harihara (Hari = Vishnu, Hara = Shiva).[12]

In these texts and icons, principles of attraction and fusion take primacy over gender. This happens not just in written texts but also in some social practices, where same-sex marriage between a God and a human is institutionalized.

Transgendered Marriage to God: *Hijra* Male-Male Weddings

Although many *hijras* are Muslims and the group probably originates from the medieval Islamicate, wherein castrated men or eunuchs guarded and served in the women's quarters of royal and noble households, *hijras* now practice an eclectic mix of religious observances. Some claim a historical link to the category of the "third sex" in ancient texts like the *Kamasutra*, a link some scholars accept with little other evidence.[13]

Hijra worship practices celebrate both chastity and sexual activity. Thus, they worship Goddess Bahuchara, who is said to have cut off her breasts to preserve her chastity from a would-be rapist. According to the legend, she asked a male devotee to cut off his genitals and dress like a woman, a practice many *hijras* emulate.

Hijras also participate in an annual celebration of a male-male wedding. The wedding is based on a legend unique to Tamil versions of the ancient epic, *Mahabharata*. Aravan, son of hero Arjuna, offers himself as a sacrifice to Goddess Kali to ensure victory for the Pandavas, the five brothers who are heroes of the epic. He asks for three boons before he dies; one is that he should be married for the last night of his life. No parent is willing to marry a daughter to him, as she would be widowed the next day. So Krishna takes a female form as the enchanting Mohini, and marries Aravan for a night.[14]

Every April, on the night of the full moon, hundreds of *hijras* congregate at Koovagam, a village in Tamil Nadu, to wed Aravan. The *hijras* dress as brides for the ceremony in a temple, which is followed by singing, dancing, and feasting. The next day, the image of Aravan is symbolically killed, and the *hijras* lament his death. Like orthodox widows, they break their bangles and other signs of marriage, and change into widows' white clothing.

As cross-dressed brides, *hijras* identify not with the human Aravan but with the divine Krishna. A traditional widow's life is supposed to lose meaning when her husband dies. Krishna, though widowed when Aravan dies, continues to live a normal life because he is a male who temporarily assumes the form of a woman. *Hijras*, like Krishna on that one night, are transgendered males, even though, unlike Krishna, they permanently take on the form of women. Their ambiguous status allows them to aspire to the sort of freedom with regard to gender and sexuality that the Gods enjoy. Like Krishna, they enjoy being brides for a night, suffer being widows for a day, and then return to their everyday lives, which often involve liaisons and marriages with men as well as prostitution. One journalist declares censoriously, "It is inexcusable that men with normal intelligence and physique be permitted to indulge in a useless lifestyle in the name of religion."[15]

Female-Female Weddings: Goddess as Husband

A parallel female tradition is that of *devadasis* (literally, God's maids), women of certain "lower" castes who worked at temples as ritual attendants, singers, and dancers, generally lived in matrilineal communities, and were known by different names in different regions.

Dedicated to Gods or Goddesses, they were viewed as embodiments of Goddesses and therefore auspicious. Like *hijras*, their presence was required at wedding and other life cycle ceremonies in all families, including "upper" caste families. Married to immortal deities, they could not be widowed, hence in some communities they were asked to tie every bride's marriage necklace.[16] Unlike most women, they inherited parental property and had considerable autonomy. One scholar reports that at the ceremony of a girl's wedding to the Goddess, the priest would warn her parents and brothers that the girl would claim an equal share in family property. They had to agree to this condition. The girl had to promise to lead a life of altruistic piety.[17]

Traditionally, *devadasis*, like *hijras*, maintained liaisons with men, often a lifelong liaison with one man. This practice (along with polygamy, homosexuality, and icon worship) disgusted British administrators. Indian social reformers and nationalists imbibed British attitudes and started a campaign in the nineteenth century to eradicate the *devadasi* tradition. In many places the practice went underground and degenerated into regular prostitution, also practiced by many *hijras*. Yet, *devadasis* suffer less social stigma than other prostitutes, since they have sacred status and are often accepted in their families and communities.[18] The battle to eradicate the practice continues today. Many governmental and nongovernmental reform agencies seize the girl children of *devadasi* families, put them in institutions, and marry them to men without their families' participation. To save their daughters from such "rescue," many *devadasi* families now marry them off as children.[19]

For our purposes, what is interesting is the practice of women marrying a Goddess. In parts of Karnataka, south India, *devadasis* are married to Goddess Yellamma in a formal wedding ceremony on full moon night in January. One *devadasi* told reporter Seethalakshmi, "It is marriage without a man. We are married to the goddess and she is our husband."[20] The *devadasis'* children are considered the Goddess's children. Though these dedications are no longer allowed, many parents still bring their daughters to the temple for the annual festival. *Devadasis* are closely associated with transgendered men, known as *jogappas*, who dress and live as females. Some *devadasis* and *jogappas* live as religious mendicants.[21]

God: Neither Male Nor Female versus Both Male and Female

In Judeo-Christian-Islamic traditions, God, though conventionally referred to as male, is ultimately without gender. "God is a spirit" (John 4:24), and, like any spirit, is neither male nor female. Angels, like Raphael and Gabriel, conventionally gendered male, are also without gender or sexuality.

In mainstream Hindu philosophy, which is heavily influenced by the Advaita Vedanta (non-dualistic) school of thought, the ultimate reality transcends gender but is manifested in everything. In other schools of thought, such as *Sankhya* and Tantra, the ultimate reality is both male and female, a duality reflected in the Gods and Goddesses most Hindus worship. Each deity is thought of as encompassing both male and female. The *Puranas* eulogize the chosen deity, whether Shiva, Vishnu, or anyone else, as infinitely flexible and available—as male, female, neuter; as animal, bird, tree, jewel, river; and as present in all elements and all forms of life. Thus, a eulogy in the *Mahabharata* identifies Shiva with a series of apparently exclusive attributes: "Thou art male, thou art female, thou art neuter."[22] Like other types of difference (class, caste, color, species), gender is irrelevant for enlightened beings who perceive the divine as pervading and/or surpassing material reality.

Ecumenical Christian theologian Diana Eck considers the idea of the Holy Spirit as active mover analogous to the Hindu principle of Shakti. She points out that in the original Hebrew version of Genesis, the Spirit that moves upon the waters at creation is *ruach*, gendered feminine, which was translated into the Greek *pneuma*, gendered neuter, and finally masculinized in the Latin *spiritus*. She thinks of the Spirit as feminine energy.[23]

To say that God is neither male nor female may appear to be not very different from saying that God is male, female, and neuter. However, there is an important difference of emphasis here between Hindu and Judeo-Christian-Islamic traditions. Absence of gender creates a lacuna, which is filled, in worship practice, by the dominant convention of maleness. Conversely, the presence of three genders (male, female, neuter) allows God to be visualized and worshiped as male, female, or androgyne. Hindu divinities have not only gender but also sexuality, and this gender and sexuality is fluid. The fluidity allows human gender and sexuality to also be conceived of as ultimately fluid even if socially constricted.

Human Gender: Ultimately Unreal

To return to Sushila's question at the beginning of this chapter, do Hindus think of the spirit as gendered? The answer, as in Christian tradition, is both yes and no. In an ultimate philosophical sense, the spirit is not gendered, but in narrative, it generally functions as if gendered. In the *Ashtavakra Gita* (ca. AD 500), quoted in the epigraph to this chapter, King Janaka learns from sage Ashtavakra that true liberation consists in recognizing the unreality of all apparent differences, including differences between men and women. But in stories of rebirth, both Hindu and Buddhist, although people are reborn in different castes, classes, professions, and even species, they are almost never reborn a different gender. Gender, in Hindu narrative, appears to be a harder boundary to cross than the species boundary. There are a few important exceptions to this tendency.

Stories where a person is reborn in another gender tend to illustrate the point that the spirit is not gendered. For example, King Puranjana, due to excessive attachment to his wife and children, is reborn as a woman, Vaidarbhi. When Vaidarbhi's husband dies, she is about to burn herself with his corpse, when a sage appears and reminds her that the true Self is neither male nor female. Here, the king represents the individual self and the sage the universal Self, which are identical. The sage says: "It is really the illusion created by me that you regard the man (Puranjana in the previous birth) as the virtuous woman (Vaidarbhi in this birth). You are neither." (IV. 28. 61)[24]

It is the female, Vaidarbhi, not the male, Puranjana, who achieves the realization that gender is ultimately unreal, and she does so as she is about to burn herself alive. Perhaps the text suggests that women need this realization more, as they suffer more of the negative social consequences of gender definitions.

So also, in the *Mahabharata*, it is the archetypal female ascetic, Sulabha, who proves that gender is unreal and women should not be socially constrained by it, since there is no real difference between a man and a woman (*Shanti Parva*, X: 65).[25] When Sulabha appears in the court of philosopher king Janaka, he objects to her communicating with him in public, as an equal. He considers such behavior improper for a woman. He asks her to which man she belongs, and what her caste is.

Sulabha, in a learned discourse, demonstrates that the Self (Atman) is the same in all beings, and is changeless while the physical and mental identity of any individual constantly changes like a flickering flame. No being can be fully separated from any other: "As lac and wood, as grains of dust and drops of water, exist commingled when brought together, even so are the existences of all creatures" (X: 65). Since the same Self is in both her and him, there is ultimately no difference between them and no impropriety in their communicating, nor does she belong to any man, as the Self cannot be possessed by anyone.

Sulabha uses the idea of the Self's genderlessness to argue for the social equality and freedom of women. She states that she is unmarried, and, in contrast to many other Hindu texts that insist on marriage for all women, here her singleness is endorsed. Arguing that a truly wise person, who has realized the oneness of the Spirit, will not try to judge anyone, including any woman, by caste or marital status, she shows that Janaka has not really attained wisdom. Janaka is silenced by her arguments, which shows that she wins this debate about gender.

Female to Male Sex-Change

If gender is unreal, why must one marry only a person of the other gender? Several ancient stories suggest that if one desires to marry a person of the same sex, one must change one's own sex first. In an ancient Greek myth recounted by Roman poet Ovid (43 BC–AD 17), Iphis is a girl raised as a boy, who falls in love with another girl, Ianthe, and laments that this love is "impossible" to fulfill. Taking pity on her, the gods change her into a man, who marries Ianthe.[26] Similarly, the *Skanda Purana* tells a story about two young men, Sumedha and Somavan, who are inseparable friends and marry each other after Somavan is transformed into a woman named Samavati.[27]

The weddings discussed in chapters 1 and 2 are represented in the media as female-female weddings even though several involve some type of behavior that may be classified as transgendered, and a few even involve a woman undergoing sex-reassignment surgery to become a man.

These cases mirror the famous case of princess Sikhandini in the *Mahabharata*. Her father, who longs for a son to kill his enemy, is disappointed when a daughter is born, and raises her as a boy. When she grows up he marries her to another princess. When the bride discovers that the groom is a woman, she complains to her parents, and a war is about to ensue. Sikhandini, grieved and ashamed, retreats to the forest to commit suicide. There, a male Yaksha, pitying her, temporarily exchanges his sex with her—Sikhandini thus becomes Sikhandin, remains married to her/his bride, and averts war. It is important to note that her marriage as a woman to a woman remains valid. As a man, Sikhandin does not remarry his/her bride.

But the sex-exchange does not escape censure. God of wealth, Kubera, is annoyed when he discovers that the Yaksha has exchanged his superior maleness for inferior femaleness. He declares, "Seeing that you have humiliated all the Yakshas by giving Sikhandin your attributes and taking the attributes of womanhood from her, and seeing that you, of wicked intellect, have done something which was never done before, you henceforth will be a woman and he a man."[28] Kubera's followers plead for a limit to the curse, so he agrees that when Sikhandin dies, the Yaksha will become male again. Sikhandin thus lives out his life as a man.

Social opinion remains divided, however. Sikhandin functions in all respects as a male. He becomes a famous warrior and even begets children. Yet his father's enemy, Bhishma, refuses to fight him, saying he will not fight a woman or one who was once a woman. Ironically, Bhishma's opinion appears to have triumphed, for in modern India, "Sikhandin" has become a pejorative term used to accuse a man of effeminacy or a person of being a fraud. The word Sikhandin literally means "peacock," which happens to be India's national bird.

Transgendered and transsexual experiences in Indian social settings, ranging from mystic and devotee to *hijra* communities, are often explained through the idea of rebirth. Transsexuals in the West report that they experience themselves as beings of one sex trapped in bodies of the other sex. This idea is familiar to most Indians, because it connects to the idea of rebirth. Powerful emotions such as love or anger cause attachments that survive death and may be intense enough to trigger a sex change. In her former birth, Sikhandini was a woman, Amba, who longed to kill her enemy, Bhishma, but could not because she was a woman. She therefore obtained a divine boon that she would be a man in her next birth. That is why Amba was reborn as Sikhandini who changed into a man, Sikhandin.

Transsexuals and Social Opinion

Modern transsexuals do not always acquire the social status of their acquired sex. Manju's family accepted her sex change from woman to man (see chapter 2); this may have something to do with their ability to absorb a son into the family business. After the sex-change surgery, Manish dropped out of the women's college where s/he had been studying, and joined the business.

Since this couple lived in New Delhi and were students at the elite Jesus and Mary College for Women, they had access to different opinions. Family friends who had traveled abroad encouraged Manju to have the surgery, telling her it is common in foreign countries. But not everyone was convinced. In a jaundiced article, Devika Rani writes that transsexuals in the West "lead a frustrated life" because they can neither have orgasms nor produce children, and that many of them want to revert to their original sex. On the basis of this unproven claim, and without interviewing the couple, she concludes that "it seems Manish has not been able to consummate the marriage," and that therefore he and his wife could not possibly be happy.[29] This narrow view of penile penetration as constituting consummation and sexual happiness is reinforced by the article's title, "Sex change has not made Manish happy: Manhood Problems," a claim not backed up by any evidence and directly contradicted by accompanying photographs, one of which shows the couple in a conventional wedding portrait with Manish's parents, while another shows them in their kitchen, playfully reaching for pots and pans. The author also claims that the doctor who performed the surgery received scores of calls from parents anxious to turn their daughters into sons. However, since the article was poorly researched and the author unabashedly hostile to gay people and transsexuals, it is not clear how much credence one should give these allegations.

One may tentatively conclude that in the Indian middle class today, a sex change from female to male is relatively more acceptable than a change from male to female, because the former is perceived as enabling social mobility upward whereas the latter is perceived as entailing downward mobility. Most *hijras* are from low-income families, many of whom reject them after they become *hijras*.

Sex-Change as Site for Same-Sex Love

In contrast to middle class Hindu society today, where most sex changes seem to be from female to male, in premodern Hindu texts most sex changes are male to female. This contradicts God Kubera's view, cited earlier, that a man would never wish to become a woman because males are superior to females. Hindu texts and religious practices abound in male to female sex changes, ranging from temporary to permanent. In some cases, the change is physical (what today would be called transsexual); in others, it involves living as the other gender without making physical changes (what today would be called transgender).

Gods as well as humans may temporarily become female. Sometimes the change results from a curse or functions as a test, but sometimes it occurs purely for erotic purposes. In Valmiki's *Ramayana*, King Ila, while wandering in the forest, enters a grove where Shiva and Parvati are engaged in love play. Shiva turns into a female to please Parvati. Mirroring him, every other being in the forest turns female too.[30] This suggests that satisfying love play includes both the heteroerotic and the homoerotic.

Because of Shiva's sex change, Ila too gets transformed into a beautiful woman, all his followers become women, and his horse becomes a mare. As a woman, Ila unites with Budha or Mercury, son of the moon. Shiva and Parvati modify the curse by making Ila a *kimpurusha* (literally, "whatman"), who is a man for one month and a woman the next. Ila produces children both as a man and as a woman. Ila finally regains manhood by performing the Ashwamedha sacrifice. In another version of the story, Ila, when of one sex, does not remember that s/he ever belonged to the other sex.

Male-Male Love as Model of Devotion

Both in Christianity and in Krishna-worship, the primary ancient model seems to be that of a male devotee beloved by a male God. Although God chooses the Virgin Mary, and Jesus has devoted female followers such as Mary Magdalene, Jesus clearly prioritizes his male followers. He appoints only males into his inner circle of apostles and reveals to them his secrets, such as the esoteric meaning of his parables and prophecies. John is repeatedly characterized as "the disciple whom Jesus loved," and Peter declared the rock on which the church will be built.

Similarly, in the *BhagvadGita*, the source text of *bhakti* or loving devotion to God, Arjuna is the model devotee. The *BhagvadGita* is part of the *Mahabharata*, which is structured around the loving friendship of Krishna and Arjuna. Krishna and Arjuna are repeatedly termed inseparable, and frequently state that their friendship is more important to them than any other relationship, including relationships with kinsmen, wives, and children.[31] Krishna tells Arjuna: "You are mine and I am yours, and all that is mine is yours also" (*Vana Parva,* XI). Krishna refers to Arjuna as half of himself, and says that half of his (Krishna's body) is made up of half (Arjuna's) body.

Arjuna asks Krishna to forgive any errors he may have committed and to bear his faults "as a father his son's, a friend his friend's, a lover [gendered masculine] his beloved's [gendered masculine]" (*Piteva putrasya sakheva sakhyuhu/Priyaha Priyaayaarhasi deva sodhum.*) (*Bhishma Parva,* XXXVI: 82).

Finally, when all the royal males are wiped out in battle, Krishna revives Arjuna's grandson, Parikshit, slain as an embryo in the womb. This child is also Krishna's nephew, as Arjuna had married Krishna's sister. He thus represents both Arjuna and Krishna, and also embodies the love between them.[32] Krishna revives the child by invoking his own good deeds, primarily the true love between him and Arjuna: "Never hath a misunderstanding arisen between me and my friend Vijaya [Arjuna]. May this child revive by that truth!" (*Aswamedha Parva,* LXVIII).[33]

Male Devotee as Bride: The Medieval Model

Through bridal mysticism, where the devotee is figured as the bride of a male God, medieval devotion, both Hindu and Christian, eroticizes the male devotee-male God dyad, and simultaneously heterosexualizes it by having the male devotee identify as female. But this heterosexualization remains incomplete because the male devotee's maleness generally coexists with his assumed femaleness. In bridal mysticism, sacred love gets expressed in erotic terms—in Christian mysticism, these terms are often drawn from the Song of Songs in the Hebrew Bible; in Hindu mysticism, they derive from ancient love poetry. When the devotee is female, bridal mysticism is easily framed in a heterosexual framework—the woman devotee identifies as the bride of Christ or of Krishna or Shiva, as the case may be. However, male devotees also identify as brides longing for the bridegroom.

In Islamic Sufi mysticism, which flourished in medieval India and continues to exert powerful influence over both Muslims and Hindus today, male devotion to the male God acquires an erotic tinge, with love for a human male beloved becoming a step to love of God.

Medieval devotion in Western Europe is feminized through the Virgin Mary, chief of saints, and the large canon of female saints, placed on equal footing with male saints. On the one hand, these saints' relation to Christ is nonsexual; on the other, it is pervaded by eroticism, in such recurrent images as that of St. Catherine's mystic marriage to Christ, where he places a wedding ring on her finger. When medieval and Renaissance male mystics identify themselves as brides of Christ, the language they use is often intensely erotic, referring to embraces, kisses, interpenetration, fusion, and burning or drowning ecstasies that irresistibly resonate with orgasmic experience. A large body of writing analyzes the hetero and homoerotic dimensions of this language.[34] However, since Christ is a virgin, as is his mother, and celibacy was the ideal of the Church, this language rarely goes so far as to refer to sexual intercourse.

This absence constitutes a crucial difference between Hindu and Christian mysticism. Hindu devotional texts, from the songs of Tamil woman poet Andal, to those of Kannada woman poet Mahadeviakka, to the Radha-Krishna mysticism of the Puranas, explicitly describe sexual embraces. Commentators often insist on the purely metaphorical significance of these sexual activities. Without challenging this interpretation, one can still argue that these detailed metaphors reflect a positive view of sexual intercourse as a worthy metaphor for spiritual communication, a metaphor also available in some texts for same-sex union.

Males Desiring Krishna, Reborn as Females

Many medieval Hindu male mystics are believed to be reincarnations of Krishna's female lovers. While Shankaracharya, the eighth century exponent of Advaita (non-dualistic) philosophy, emphasized the importance of realizing the identity of all souls with the one universal soul, Ramanuja thought this realization was only a stage in the journey toward the goal of love between individual spirits and the universal spirit or God.[35]

The *Padma Purana* shows sages performing austerities not to become indistinguishable from God but to become eternal females engaged in amorous sports with God. Here, men contemplate an erotic form of the male. For instance, one sage meditates on Krishna "lying on his back on the beautiful bed of leaves, whose expansive chest was being repeatedly covered by a cowherdess, who was greatly overcome with passion and whose eyes were red, with her breasts, who [the lord] was being kissed on his cheeks, and whose lips were being gratified, who, the wonderful one, was with a smile holding his beloved with his arms."(V. 72. 12–19a).[36] Another meditates on "Krishna who was of the form of joy, who was moving along the streets of Braja with a strange and sporting gait, who was making a jingling sound of his anklets with charming steps, who attracted the minds and bodies of the beautiful women of Braja with the knots of their garments loose and suddenly embracing him . . . " (V. 72. 34–46).

These sages are reborn as daughters of various cowherds, and become the eternal lovers of Krishna. The text lists the names of the sages as well as of the women they become. There is no physical description of the male sages—we are told only about their knowledge and purity, and the penances they perform. But when they are reborn as women, their physical beauty is described in detail. When male becomes female, abstract becomes concrete, mental becomes physical, and sacred chastity becomes sacred eroticism.

While these texts show males permanently becoming females, others show a temporary sex change enabling sexual union. Such are the stories of divine sage Narada and of Krishna's friend, Arjuna, consummating their devotion to Krishna by temporarily becoming his female lovers.

Narada seeks to know Krishna's secrets. Vishnu arranges for him to bathe in a lake filled with divine nectar. Emerging from the lake, Narada finds that he has become a woman, and engages in love sports with Krishna for a year. Krishna says, "I am truly of a feminine form, and I am the ancient woman, and I am goddess Lalita, and in a manly form I have Krishna's body. O Narada, there is no difference between us" (V. 75. 45). This suggests not only that the divine male and female principles are indistinguishable but also that the ascetic and devotional traditions are inseparable (Narada and Krishna are one).

This makes it possible for the *Padma Purana* both to condemn women's unlicensed sexuality, and also to exalt the polyamorous behavior of Krishna with numerous divine damsels. The latter is understood as symbolic, and thus the divide between flesh and spirit remains. My contention, however, is that, despite the condemnation, the detailed and extensive celebration of these amours has some effect on discourse about human sexual relations.

Arjuna as Woman: Friend Becomes Lover

In representing Arjuna temporarily changed into Krishna's female lover, the *Padma Purana* adds a significant sequel to the *BhagvadGita*. In the *Gita*, Arjuna figured himself as Krishna's male beloved, but in the *Padma Purana* he can be Krishna's beloved in the fullest sense only if he turns into a woman. Instructed by divine damsels, Arjuna worships a Goddess and bathes in a lake. He emerges as a woman with no memory of her/himself as a man. After she worships the Goddess again, the new woman is termed Arjuni. The text also terms her "Arjuniya," an affectionate diminutive. These names are significant, because the female names keep alive for the reader the famous male name. In hearing this cross-sex love story, the devotee is not allowed to totally forget that this is also a same-sex love story.

Arjuni is taken into a pleasure grove whose beauties are elaborately described, in accordance with Sanskrit romantic convention. There s/he meets Krishna who secretly sports with her. Arjuni is then made to take another bath and turns into Arjuna once more.

Arjuna remembers his female incarnation, and feels "depressed and heartbroken."[37] The reasons for this dejection are not explained—it could be the typical mystic's depression on returning to earthly life after a brief vision. Arjuna may wish he could have remained Arjuni forever. The depression lifts when Krishna reassures him, and we are told that Arjuna ultimately goes to Krishna's eternal abode, and remains there, knowing the sports of Krishna. We are not told whether he does so in the form of Arjuna or Arjuni.

Sex-Change and Rebirth

The *Padma Purana* narrates many stories of men actively willing themselves to be reborn as women, in order to experience erotic bliss with a male God. For example, Chitradhvaja, twelve-year-old son of a royal sage, goes alone to the temple of Vishnu and meditates on Krishna. He has a dream, in which Krishna's female lover, at Krishna's command, turns Chitradhvaja into a girl. She does this by thinking about his body as non-different from hers. As a girl, Chitradhvaja is named Chitrakala and has "charming round hips" and full breasts, and becomes the servant of Krishna's female lover.[38] When embraced by Krishna in the dream, the boy Chitradhvaja wakes up. He then behaves like the standard woebegone lover. He gives up food and pleasures, does not speak, and weeps constantly. He goes to the forest and practices penances, which result in his dying and being reborn as Chitrakala, one of Krishna's beloved milkmaids.

The *Padma Purana* interprets these visions and rebirths as symbolic, not literal. A sage, who has a vision of Krishna, asks him the meaning of these symbols. Krishna tells him that the cowherd men should be understood as sages, the cowherd women as the Vedas, the young daughters of the cowherds as hymns, and so on. Nevertheless I would argue that this mystical significance, since it is couched in erotic symbols, has an overflow effect, endowing human eroticism with positive meaning for the devotee.

Sex-Change and Same-Sex Desire

These stories suggest historical shifts in the ways sex-change and its relation to desire are represented and understood. In the ancient texts, a permanent change from female to male, like that of Sikhandin, is seen as good for the individual concerned, but not everyone accepts its reality. The change enables same-sex desire to be consummated in the form of cross-sex desire. A temporary or continually reversible sex-change like that of Ila is not perceived as entirely good. It too enables same-sex desire to be consummated in the form of cross-sex desire.

In medieval texts, both permanent and temporary sex-changes are good (Somavan to Samavati; Vishnu to Mohini; Arjuna to Arjuni; Narada and the other sages). They enable same-sex desire to be consummated in the form of cross-sex desire. In modern India, permanent sex-change from female to male is viewed as better than permanent sex-change from male to female. Neither is socially perceived as entirely "real," and both enable same-sex desire to be consummated in the form of ostensibly cross-sex desire.

What all of these sex-changes have in common is that they make same-sex desire more socially acceptable by rewriting it as cross-sex desire. The only sex-change that enables same-sex desire in itself is that of Shiva who temporarily becomes a woman to please his wife. All of these patterns of sex-change appear in stories frequently rewritten and retold in India, in oral and folk literatures as well as written texts. The Sikhandin story has been particularly fruitful in this regard. But the only Sikhandin type text I have come across that enables same-sex desire to be consummated in marriage without a sex-change is Vijay Dan Detha's version.

Choosing to Be Women Together

Dan Detha is a major writer of modern Rajasthani fiction, some of whose stories are inspired rewritings of folk tales. Like Shakespeare's rewritings of popular tales, Dan Detha's narratives involve women cross-dressing, passing themselves off as men, and getting romantically entangled with other women. Detha's story, "A Double Life" (*Dohri Joon*) was translated into Hindi and performed as a play "*Beeja Teeja*," in New Delhi in the early 1980s. It appeared in Hindi under the title "*Naya Gharvas*" (A New Domesticity), and I first translated it into English in *Manushi* No. 17, 1983.[39]

This story of love between two women, Beeja and Teeja, is written in lyrical, alliterative Rajasthani that combines the pleasures of oral literature with those of literary texts. It draws on earlier motifs but gives them a surprising twist—here, the woman who is miraculously enabled to become a man realizes that she prefers to be a woman and changes back into a woman, yet remains married to her female lover.

The title is not easily translatable into English. *Dohri* can mean double but also dual, reduplicated, or twice as much. It has the same root as *dogana*, used to refer to female-female lovers in *rekhti* poetry (see chapter 8). It suggests human or divine potential to be both male and female as also to double oneself through love for another. *Joon* comes from the root for womb/vulva (Sanskrit *yoni*), and thus refers both to birth and to the body assumed by the soul at birth. Indian-language words for birth often suggest not a one-time event but a series of births and lives. The title thus alludes to the tradition of sex change occurring through rebirth. Finally, since *joon* refers to the female sex organs, the title "Double Yoni" refers to a sexual relationship between two women (see chapter 6 for a similar formulation).

The story begins by invoking Kama, and asking that all listeners be granted two lives or a double life. The story is set in motion by the Sikhandin pattern—a girl child (Beeja) is raised by her father as a boy and married off by him to another girl (Teeja).

But after the wedding, when the two young women discover the deception, the pattern changes. While Sikhandini's bride is upset and wants to end the marriage, Beeja's bride insists that they should stay married, and live together publicly. This evokes extreme hostility from the groom's family and village.

The tragic ending of some lesbian love stories in India today is averted here by the intervention of a ghost, who appears to the women in the forest where they flee to escape the villagers' censure. The forest, in Indic as in Indo-European narrative, is a liminal place of secrets and transformations; it may be dangerous but is also miraculous. Just like the forest spirit who enables the suicidal Sikhandini to become a man, this ghost helps the two women. He gives them a magical palace in the forest where they live happily. Women can visit them but not men. The villagers are too scared of the ghost to further persecute the girls.

The palace is described as a female world, suffused with rosy light: "Saffron courtyards. Crimson walls. Vermilion ceilings. A lotus bedstead. A bed of roses. They swung in the swings of joy."[40] The narrator contrasts the two women's bliss with the domestic discord, gender inequality, and property disputes prevailing in the village. The women's home serves as a refuge for oppressed village women.

But when Beeja realizes that the ghost has miraculous powers, she requests to be turned into a man. This sex-change constitutes the "happy ending" of the original Sikhandin story. Here, however, it is differently framed. Teeja is opposed to the idea—she feels they have already reached the pinnacle of happiness and nothing could make them happier. The ghost seems to agree, for when he grants Beeja's desire, he adds that if she ever desires to become a woman once more, she will be able to do so.

Beeja, who was raised as a boy, is thrilled when she becomes a man, but his behavior now changes and he starts to bully Teeja. Beeja wants to be acknowledged as the sole owner of the palace, aspires to accumulate more property and many wives, forbids Teeja to go out alone, and even accuses her of infidelity with the ghost. A quarrel results, in which Beeja hits Teeja, and then runs out alone. When he reaches the mountain peak where the two women first consummated their love, he remembers their happiness and wishes to be a woman again. He is immediately transformed into a woman, and the two live happily ever after.

The story follows medieval narratives, like that of Arjuna in the *Padma Purana*, in its celebration of the joys of men becoming women, but it removes the male God from the equation. It also introduces an unsettlingly radical analysis of male oppression of women.

Sex and the Spirit

The story is full of utopian descriptions of the two women's lovemaking. Their desire is identified with the primal Kama and also with the whole world's desire, in a manner reminiscent of sacred medieval texts where Radha and Krishna's or Shiva and Parvati's lovemaking becomes a metaphor for the play of the universe:

> "At dawn when they came out of the palace and saw the sun rise, they felt as if the sun were rising from the pure petals between their thighs. Ever since that night, the sun has forsaken its former dwelling and has begun to rise from this new abode, whence it rises even today. All the joys of the world throbbed with eagerness to dwell in the bed of that palace. The thirst of the whole universe was encompassed in that one thirst of theirs."[41]

However, after Beeja becomes a man, the same set of symbols is used to show male power developing. The story suggests that pride arises in Beeja primarily from the experience of heterosexual sex:

> "He picked Teeja up in his arms. All of her struggles were of no avail. Laying her down on the bed of roses, he fell on her. . . . The petals of the lotus seemed about to break asunder but did not. . . . a new knowledge began to dawn in the husband's mind—that a man is stronger than a woman. . . . A man is indeed tremendously powerful. . . . When the husband's eyes opened, the sun had already climbed into the sky. . . . Seeing the rays, pride awoke in his heart, telling him that it is man's heat and power that rises in the heavens in the form of the sun. Woman is merely his shadow."[42]

I suggest that this functions as a critique not of heterosexual sex or manhood in general but of the dominant obsession with penetrative sex. Beeja's focus on penetrative sex leads him to a comically lopsided view of his place in the world.

Same-Sex Marriage and the Ghost of Tradition

This story celebrates the marriage of two women, much as some families in India have recently celebrated their daughters' marriages to other women. The women's love story follows the classic trajectory of Indian romance narrative—lovers suffer, defy society, and act for the good of others, not just themselves. The story incorporates many tropes from romantic convention, such as the monsoon, sun and moon, lotus and rose, forest and garden.

The ghost constitutes the surprise factor that makes the story more than a feminist tract. For one thing, his maleness is surprising. While all other males in the story are bullies and cowards, he is chief and spokesman of a troop of ghosts who haunt the forest, and are collectively hostile to the censorious villagers but delighted by Beeja and Teeja's defiance of patriarchal norms.[43] I suggest that the ghosts collectively stand for literary traditions wherein male writers sympathetic to women have written pro-women texts based to some extent on their interactions with women friends and family members (see chapter 8).[44] In the story's concluding lines, the narrator remarks: "I wrote this story at Teeja's dictation, in her words. Would the ghost chieftain have spared my life if I had dared add a word to her account?"[45]

To be a ghost is to be less than real but also more than real. Like the bodiless Kama, the ghost represents a truth superior to fact. The convention-bound villagers are haunted by him and cannot wish him away as unreal.

Just as Hamlet's father's ghost stands for values swept aside by the advent of modernity, the ghost in Detha's story may stand for premodern, precolonial traditions, sympathetic to desires outlawed in the postcolonial present. As a ghost, he is literally a spirit. He may be read metaphorically as the spirit of the literary and imaginative tradition that makes possible the seemingly impossible, enabling, for instance, some women today to successfully invoke tradition when they marry one another.

Chapter 4

"Immortal Longings": Love-Death, Rebirth, and Union Through Life After Life

Destiny never considers whether a union is possible or impossible.

—Kathasaritsagara *(11th Century AD)*

A pair of star cross'd lovers take their life . . .

—Romeo and Juliet, *Prologue*

They do not think there can be tears between men.

—*Christopher Isherwood,* A Single Man, *(1964).*

If marriage is a public statement of a couple's intent to live together, joint suicide or love death, wherein a couple express their commitment by dying together, can also function as a type of marriage—a public statement of intent to unite forever. This "forever" may be conceived of as a real site—the next world, the afterlife, or future lives. It may also be the "forever" of memory and history—death may make visible and public on this earth a love that society would not permit to be consummated in marriage.

Ironically, it is in death that what is common to cross-sex and same-sex relationships becomes most starkly evident. Every year, many Indian couples, both cross-sex and same-sex, commit joint suicide. Very similar patterns are evident in these suicides. The many same-sex couples, most of them women, who have jointly committed suicide in the Indian subcontinent in recent years have inscribed themselves into a generally understood pattern of love and marriage, while simultaneously putting themselves beyond social and state interference.[1]

In the United States, it is estimated that of every three youths who commit suicide, one is lesbian, gay, bisexual, or transgendered. This suggests that LGBT youth commit suicide in disproportionately large numbers. These suicides are generally seen as the result of a young person suffering extreme isolation and/or persecution as a result of his/her homosexual identity or feelings. Some of these suicides are responses to enforced separation from lovers. Enforced heterosexual marriage is not as common in the West as in India. But families' attempts to enforce a heterosexual identity through other means, such as conversion therapy, prayer therapy, and societal conventions, such as the pressure on teens to date the opposite sex, have similar emotional effects.[2] Some families throw out gay children, who end up homeless. The fear of such disinheritance also drives some to suicide.

In India, families routinely arrange the marriages of young people, and when parents discover a youth's same-sex involvement they often respond by hastening the arrangement of a heterosexual marriage. Same-sex couples' suicides in India are similar to individual LGBT suicides in the United States insofar as both are a response to compulsory heterosexuality.

All the couples, cross-sex and same-sex, who commit joint suicide in India, are under severe emotional stress. They are subjected to social and familial pressure to separate from their lovers and marry others, and are also often in physical fear of being injured or murdered by hostile relatives.

In family negotiations and conflicts concerning marriage, parents and children in India often hold out the threat of suicide to compel the other to give in. This is a standard aspect of the "family pressure" to marry that many young people experience, and also of the counter-pressure they exert. In 2003, Sheela, 23, and Sree Nandu, 21, used this threat to assert their right to live together (see photo 4.1). Sheela, who had a child out of wedlock, was abused and imprisoned by her family. She and Sree pretended to consume poison, and then escaped from hospital together Sheela dressed as a man, and they lived as husband and wife in Waynadu, Kerala, for two months. When a local tabloid exposed and defamed them, the police threatened them with prosecution under the antisodomy law, and insisted that Sheela return to her father. With the help of some activists, they convened a press conference, where they stated: "We will live together till our death. No force on earth can separate us. If society and you press people don't allow us to live as lovers, we have no option but to commit suicide."[3] This turned the tide in their favor, and they found shelter with a social welfare organization in Bangalore. They have now founded an organization to help women who want to live with each other.

Couple suicides appeal to three widely prevalent ideas—love as protest against social injustice, the inevitability of lovers' destiny, and lovers'

Photo 4.1 Sheela (foreground) and Sree Nandu. Photo: *Savvy*.

reunion in the afterlife. Death constitutes the end of an individual life, but its social and cultural meaning may exceed that finality. The choice to die may be represented and understood as a choice to live on in other ways, a choice of immortality.

Couple suicide may have an impact on public attitudes to love. At the end of Shakespeare's *Romeo and Juliet* (ca. 1595), among the most famous stories of love-death, the ruler of the city tells the fathers of the dead couple, "See what a scourge is laid upon your hate, /That heaven finds means to kill your joys with love" (V.iii.292–293).

Joint Suicide as a Rite of Marriage

Suicide is an act simultaneously intensely private and public. Committed in private, it is nevertheless public because it draws the scrutiny of society and state. Suicides by same-sex couples are generally enacted as public statements. Many couples leave behind written accounts of their motives, and express last wishes. Some secretly marry before they commit suicide. Many families nevertheless manage to cover up these suicides— for every reported death, many more probably go unreported. In some cases, however, the couples achieve in death what they could not in life—a public acknowledgment of their commitment. Such a statement and acknowledgment in life is usually termed marriage. In that sense, these suicides to some extent function as same-sex marriages, or, in cases where secret marriage preceded suicide, function to declare those marriages.

The Christian wedding ceremony requires spouses to vow fidelity "until death do us part." Christ famously remarked that marriage does not carry over from this life to the next, since in heaven there is neither marrying nor giving in marriage. But European Christian culture developed other ideas. Spouses such as Romeo and Juliet, deprived of an extended married life on earth, are immortalized in works of art and remembered as wedded in death. Lovers who are unable to wed in life are also remembered as wedded in death. Such are Paolo and Francesca, whom Dante, in his *Inferno*, sympathetically represents as preferring union in hell to separation in heaven.

In *Antony and Cleopatra* (ca. 1607), Shakespeare shows Antony committing suicide to reunite with Cleopatra, who he wrongly thinks is dead. After he dies, Cleopatra performs her suicide as a wedding ritual. Her two maids, who serve as bridesmaids, dress her like a bride, and she anticipates happiness in the next world with Antony: "I have /Immortal longings in me" (V.ii.280–281). Although Antony was married to another woman, and Cleopatra, queen of a subjugated nation, could not be his wife in this life, she sees herself as married to him in the next life. As she applies the poisonous snakes to her breast, she addresses Antony,

"Husband, I come"(V.ii. 287). Shakespeare thus has the audience witness a wedding celebrated in the act of suicide.

The young couples who commit suicide in India also perform rites that transform their deaths into marriage-like unions. For example, when Bindu, 21, and Rajni, 22, committed suicide by jumping into a granite quarry in Kerala in January 2000, they tied their bodies together with a *dupatta* (woman's scarf).[4] Tying the bride's and groom's clothes together to symbolically unite their bodies as they walk around the fire is a ritual common in Hindu weddings. A few days earlier, the two girls had tried to elope together but had been prevented from doing so. Both wrote notes to their families, saying that they were killing themselves because they had realized the impossibility of being able to live together.

Lalithambika and Mallika, with whose story I began this book, tied their hands together with a ribbon when they jumped into the Cochin channel. Joining of hands is a rite central to most wedding ceremonies. In the Hindu version of this rite, the spouses' right hands are often tied together with a red cloth. Heterosexual lovers make similar gestures; for instance, a man and woman who jumped together into the Sutlej river in December 2002 tied their arms together with a piece of cloth.[5] The woman was married to another man, and had two children.

On April 16, 1992, the *Indian Express* reported that two Nepalese women, Rekha Puri, 22, and Gayatri Parayar, 18, committed suicide together "after their families refused them permission to be married to each other." They were found hanging from a tree in their village, Rajapur. The newspaper report, entitled, "And in their death they were united," notes that the Nepalese news agency RSS reported: "It is presumed that they had informally performed the marriage rites since Gayatri was wearing the *sindoor*." The women also left behind a suicide note, in which they wrote, "we want to live as husband and wife in the life beyond the grave."[6]

Civil marriage is certified in writing; Hindu weddings do not require written certification but they generally have witnesses. Suicide can rarely have witnesses, but it does have witnesses after the fact. Through their suicide notes, the couples call upon these witnesses to posthumously ratify their unions. Through the words they write together, they make their vows public.

One element found in almost all the letters is a reassertion of the partners' refusal to be parted. This refusal gestures toward the intense harassment by family and community members that these couples experienced. But couples also often state that no one should be blamed for their deaths. Mamata and Monalisa stated in their note, written in blood, that no one should be held responsible for their deaths.[7] The paradox of love death is well expressed here, with the blood, as it were, contradicting the words.

Cross-Sex Love Suicides

In many cases, the final push toward suicide seems to be the looming threat of an unwanted marriage for one or both women. Heterosexual love suicides are also often precipitated by the woman's family pushing her into marriage with another man. In most heterosexual cases, families oppose the union because the lovers belong to different religions or castes, or one of them is already married. Some orthodox parents simply find the idea of children choosing their own spouses unacceptable. In October 2002, a man, Vikram, 22, and a woman, Suman, 17, consumed poison together in Delhi, leaving behind notes written on one sheet of paper, in which they stated that their parents opposed their marriage, but no one was to blame for their deaths.[8]

In February 2004, a man, M. Anjaneyulu and a woman, P. Mamatha, aged 19 and 17, consumed poison together in Hyderabad, because their families opposed their affair, and Mamatha's father had arranged her marriage to another man. While her parents were at the prospective bridegroom's house, conducting a ritual worship, the lovers committed suicide.[9] In December 2003, Shyam, 25, and Noorie, 18, jumped in front of a train after their families refused consent to their marriage. He died and she lost a leg.[10] In March 2004, Naresh, 22, and Veena, 21, who had been in love since their teens and plunged into depression after she was married off to another man, jumped off a high-rise building in Hyderabad.[11]

Sometimes, a love suicide seems like the only option to a couple whose marriage would be considered impossible in their community, and whose love is discovered. In 1937, a friend of M.K.Gandhi's wrote to him about a girl, 16, and her maternal uncle, 21, who were in love. In their community, such a marriage would be considered incestuous although it is permissible in some other Indian communities. When the girl became pregnant, and the affair was discovered, both of them consumed poison.[12]

Suicide Preceded by Same-Sex Marriage

In several cases, same-sex couples performed wedding rites before they committed suicide. In one of the only two reported cases of a male couple committing suicide, an 18-year-old boy and his male partner got married in 1991 and lived together for two years "as husband and wife" in Trissur, Kerala. The newspaper cryptically reports that in 1993 they stabbed each other to death because "of the non-recognition of their marriage by society."[13] The other male couple, Suresh, 19, and Krishnakumar, 17, described as "inseparable friends since childhood," consumed poison together in Shoranur, Kerala, in 1999.[14]

Mamata and Monalisa, whom I mentioned in chapter 2, registered a life partnership deed four days before their attempted suicide. This "Deed of Agreement for Partnership as well as to Remain as Life Partner," drawn up by an advocate and registered in court on October 6, 1998, is reproduced in an ABVA fact-finding report, which is one of the very few investigative reports on a same-sex joint suicide.[15] Both women signed the deed, two men witnessed it, and a public notary authenticated it. The deed is worded in legalese, which the advocate probably thought appropriate, but its emotional content becomes apparent in some sentences.

The deed provides details of the two partners' ages, parentage, occupation, and residence, and then states that both are "bachelors and have intimated their relationship with one another for last several years." It continues, "Whereas their relationship has become so close that it is not possible on the part of either party to live apart or sever such a relationship," therefore they "have decided to live together as Life Partner forming a partnership for the purpose of earning their livelihood." The meaning of "partner" as business partner is thus incorporated into the deed, but this is clearly a pretext, as the nature of the business is not specified: "On and from today the first party and the second party shall live together and by means of any business to earn their livelihood. The partnership shall be known as Mamata & Monalisa." The deed states that the partners have "a capital of Rs. 1,000 contributed equally," and that they intend to invest it in "some sort of cottage industry" to help handicapped women, widows, divorcees, destitute women, and orphans.

After this vague statement of intent, the equivalent of vows are incorporated into the deed. The partners agree to "remain bachelor" and to fulfill their ambition in life of helping the poor, and "to accept unmarried girls as partners of their temperaments." This last phrase suggests that they think there may be other girls who are temperamentally like them. They also agree not to ill-treat or annoy each other or inflict mental or physical cruelty on each other. They agree "to continue their life as Life Partner for good" and to "create an atmosphere for healthy, sound and peaceful living."

Four days later, on October 10, the two girls took poison and also stabbed each other in Mamata's bedroom. When discovered by Mamata's mother, Kamla Devi, they were both wounded and crying. Mamata told her mother, "As society did not allow us to live together so we have committed suicide," and Monalisa added, "No one will be able to part us from each other."[16]

Physical Intimacy

A third type of document left behind by some couples consists of love letters to each other, containing acknowledgments of physical intimacy, which would normally be considered shameful. Death, as it were, frees these women to state the truth about their relationships.

Gita Darji and Kishori Shah, 24-year-old nurses who hanged themselves together in 1988 in the hospital where they worked in a village in Gujarat, western India, left behind suicide notes, stating that they did not blame anybody for their action. They also left behind two sets of letters. In a letter to Gita, Kishori wrote, "I can't live and sleep without you."[17]

Mallika and Lalithambika also left behind their love letters. In a card, Mallika wrote in red ink, "here are a thousand kisses for you in public." Below this, Lalitha wrote in green ink, "Come to me. I shall take you in my arms. I shall cover you with kisses. You shall sleep on my bosom and afterwards, maybe, we shall have a little quarrel."[18]

Traditionally, weddings have been occasions when it is permissible to talk about sexual intimacy. The Church of England wedding ceremony includes the vow, "With my body I thee honor"; at Indian weddings, both men and women participate in raunchy joking, singing, dancing, and teasing of the bride and groom. Discussion of sex, impermissible in polite society in India, becomes possible at weddings. Going into death, these couples are able to make public their sexual relationship in a way that would only be possible in life if they were allowed to marry.

Suicide as Protest: The Cultural Ideal

Joint suicide by lovers or spouses is almost always a form of resistance to social forces that seek to part them. This type of suicide is part of a larger cultural ideal of suicide as a noble form of resistance to injustice. Instead of violently attacking the oppressor, the oppressed person protests by turning the violence inward, against him or herself. This ideal is found in ancient and medieval Indian texts, where Brahmans and Buddhist or Jain monks protest a tyrannical ruler's behavior by fasting to death or burning themselves alive. As historian Dharampal has shown, Gandhi drew some of his strategies, including that of the fast-unto-death, from such indigenous traditions of civil disobedience.[19] Similarly, some Vietnamese Buddhists, including a monk, immolated themselves to protest the Vietnam War.[20]

Gandhi's idea was that this type of action changes the oppressor's heart through love, and awakens bystanders to their responsibility for injustice. Such protest can also be undertaken, however, in anger and despair. For instance, in 1990, dozens of "upper" caste youth in urban India, both men and women, protesting what they saw as unjust reserved quotas in education and employment for castes somewhat lower in the ritual though not necessarily in the economic hierarchy, committed suicide by burning themselves to death in public. Couple suicides represent a mixture of love, anger, and despair, directed primarily against family.

Indian narratives often represent couple suicides resulting in reform of tyrannical parents and thus of society. In life, it sometimes but not always has this effect. Many Hindi films have idealized joint suicide by lovers as protest against parental tyranny. In *Ek Duuje ke Liye* (For One Another 1981), a south Indian boy and north Indian girl, whose families live next door to each other, fall in love. Although both are Hindus, the parents are appalled. The forcibly separated lovers pine for one another, inflict various kinds of torture on themselves, and finally commit suicide together, leaving the parents to unavailing remorse. In *Bobby* (1973), attempted suicide by the young couple changes the parents' hearts just in time. The lovers, a rich Hindu businessman's son and a poor Christian fisherman's daughter, jump into the ocean. Their remorseful fathers rescue them, and then proceed to approve the marriage.

Suicide as protest is, however, a double-edged sword. In an exceptional variation on the pattern of love-suicide, one woman killed herself in protest against her sister's proposed marriage to a woman. In 2002, in Durg, Madhya Pradesh, Anjali, 22, a nurse, began living with Dr. Neera Rajak, 39, whom she planned to marry. Anjali's father lodged a police complaint, accusing Dr. Rajak of kidnapping Anjali. Both women appeared before the police, and a judge ruled that they could live together, as they are adults. Anjali's sister Suman, 18, then committed suicide, saying she could not tolerate Anjali's "shameful activities."[21]

Forced into Marriage

Newspaper reports of same-sex couples' suicides generally dwell on the details of the deaths but do not reveal much about the social hostility the couples faced. Thus, when two young girls, Suchita Lonare and Seema Balanti, consumed poison together in a field in Mandar village, Maharashtra, western India, the paper reported that they "were very close friends who deeply loved each other" and added, "Why did they consume poison? No information could be obtained on this."[22]

In other reports, one glimpses family hostility. Thus, when the two nurses Gita and Kishori hanged themselves together in Gujarat, police found that Gita had been married to a man a few months earlier, and he "abhorred" her relationship with Kishori. He complained to Gita's brother who "made Gita apply for a transfer." The report states that the "pangs of impending separation" drove the two to commit suicide.[23]

In December 2003, a Nepalese newspaper reported that a young woman, Anjali Thapa, had attempted suicide by consuming poison because her parents were trying to separate her from her partner, Sushila Gurung, and were forcing her into marriage with a man. Even after the suicide attempt, Anjali felt unable to tell her parents the truth.[24]

Social pressures outside the family also play a role. Remaining unmarried is a viable if somewhat unconventional option for some middle-class women in India. But if such women's relationships with each other become public, fear of harassment may drive them to suicide. Sumathi, 26, and Geetalakshmi, 27, had been living at a Yoga Center in Coimbatore, Tamil Nadu, for three years. Since they were somewhat beyond the conventional age for marriage, and were living away from their parents, it appears that their religious way of life allowed them to remain respectably single. But the people at the Yoga Center found out about their intimate relationship, and threw them out. They had to part and return to their parental homes. Unable to bear the separation, they decided to commit suicide. In a letter to their guru at the Center, they wrote, "We did a mistake because of which you threw us out. . . . We cannot survive in this society. That is why we arrived at this decision. Please forgive us."[25]

In some cases, fear of police harassment is a factor. Mini, a graduate student in Trichur, was accused of having a lesbian relationship with her friend. Both girls disappeared for five days, and apparently went to Chennai. When they returned, they were sent back to their parents, but since police complaints had been lodged regarding their disappearance, Mini was asked to go to the police station the next day. Instead, her body was found floating in a reservoir, with a suicide note in her hand.[26]

Police often illegally intervene to separate same-sex couples. For instance, in April 2000, when two Nepalese girls married each other by a Hindu ceremony, the police separated them and returned them to their parents, although they had broken no laws. The girls were quoted as saying: "No one can break our nuptial chords nor make us cease to love each other."[27]

Violence: How the Family Works

Young women who resist marriage or try to marry against their parents' wishes often face violence or threats of violence. The story of two young women, Kiran and Khurshid, demonstrates the extremes to which some families go. In 1983, when I was working with the women's organization *Manushi*, in New Delhi, Kiran Shaheen, a feminist activist who ran her own printing press in Patna, Bihar, eastern India, wrote us a letter in Hindi, describing how her companion Khurshid Jahan, had been kidnapped by her family. Kiran was from a Hindu family while Khurshid was from a landowning Muslim family. Both were in their mid-twenties. Khurshid had been married to a man when she was ten years old, but had never lived with him.

Kiran and Khurshid decided to rent an apartment together. Khurshid had completed her MA and BEd and was looking for a job as a schoolteacher. Since her family opposed the idea of her living independently, she stopped taking money from them. In June 1982, the family wrote to Khurshid, saying that her mother was dying. When she returned to her village, she found that her mother was quite well. The family had plotted to send Khurshid to her husband. Then, her father discovered that Khurshid's husband had married again, so Khurshid was allowed to return to Patna.

In 1983, the family repeated the ruse, telling Khurshid that her father was ill. Khurshid left with them but did not return as promised. Extremely anxious, Kiran took a male friend, Suresh, an activist with the People's Union for Civil Liberties, and went to Khurshid's village. They found that Khurshid had been imprisoned in a room without food, abused, and threatened that both she and Kiran would be killed. The family said they planned to marry Khurshid to someone in a remote village. Khurshid's male relatives manhandled Kiran and Suresh, threatened to kill them and strip Kiran naked, and forced them at gunpoint to sign false statements. The local police colluded with Khurshid's influential family, and refused to help Kiran, saying the case could lead to Hindu–Muslim conflict because Khurshid was a Muslim.

Kiran wrote to us, "I find darkness all around me. We both had decided to live together. . . . People here cannot understand a girl's feeling of love and respect for another girl. Even other girls only laugh at such a feeling and consider it absurd. But I am sure you will understand my suffering."[28]

Kiran then came to Delhi, and, with *Manushi*'s help, filed a habeas corpus petition in the Supreme Court, asking for Khurshid's release from illegal confinement by her family, and the restitution of her constitutional rights to life and liberty. When Khurshid was produced in court, she succumbed to family pressure and said she wanted to stay with her family. But a few years later, when the drama died down, Kiran and Khurshid got back together, and moved to Delhi. Khurshid's family quietly gave up and let her go her own way.

This may appear to be strictly homophobic violence, but in fact it is not. It is violence against the assertion of independent choices. Homophobia even in the West is only the most visible dimension of a phobia against all non-normative sexuality. Kiran Shaheen had earlier come to Delhi to file a similar habeas corpus petition on behalf of Kiran Singh, a Hindu girl who was nearly killed by her family for wanting to marry a Muslim boy. Kiran Singh narrowly escaped death, fled to Delhi, and took refuge at *Manushi*. After her petition was upheld in court, she and her boyfriend were married and celebrated their wedding at the *Manushi* office.

Family Politics Continue After Death

When one partner in an attempted suicide dies, but not the other, the dead partner's family may sue the surviving partner for murder. This happened in the case of Mamata and Monalisa. Mamata survived her injuries but Monalisa died, and Monalisa's family then charged Mamata with murder.

The situation gets more confused in the context of so-called honor killings. For example, in March 2003, 16-year-old Bimla's three brothers killed her and her 22-year-old male lover, Guddan, when they were discovered together one night in their village in Uttar Pradesh. She was Muslim, he was Hindu, and the two belonged to the same village, so the affair was also seen as incestuous. Bimla's brothers confessed, without remorse, to having committed the murders to defend their honor. Bimla's family had arranged her marriage to another man; when asked if she had agreed to this arrangement, her father said, "In the village no one asks the girl."[29]

In an atmosphere fraught with familial violence, it sometimes becomes hard to distinguish suicide from murder. In December 2003, a young man and woman who were lovers were found dead in a village in Pakistan. Her father had refused them permission to marry. The police thought the couple had entered into a suicide pact but the man's family alleged that the woman's family had committed a double murder and arranged the bodies to look like a joint suicide.[30]

When a family objects to a marriage, they may continue the feud for years. This happened in the case, discussed in chapters 2 and 3, of Madhu's marriage to Manju, who became a man, Manish. Madhu's father sought legal means to stop the marriage but failed; however, he found an opportunity to vent his hostility four years later. In a rare follow-up story, papers reported that Manju had been seriously burnt, and Manish had suffered burn injuries while trying to put out the fire. The couple lived in a joint family, with Manish's parents and brother. By Manish's account, Manju's synthetic clothes accidentally caught fire when they returned from a late-night outing and she began to heat up milk on the gas stove. While such accidents do often happen, sometimes they are also wife-murders or forced suicides disguised as accidents. Madhu's father seized the opportunity to initiate a police enquiry. The photograph accompanying the report shows Manish disheveled, bandaged, and weeping.[31]

When the Just Die, Are the Unjust Punished?

In Indian legends, lovers prove the authenticity of their love by suffering ordeals. If they pass these tests, the Gods may intervene to protect them. But if society persists in rejecting the match, the lovers' righteous anger may wreak havoc.

In the seventeenth-century Punjabi romance *Heer Ranjha* by Muslim Sufi poet Waris Shah, still very popular today, after the adulterous lovers, who are Muslims, endure many ordeals and finally elope together, a judge returns Heer to her husband. The separated lovers curse the town for its injustice. Heer invokes Biblical and Islamic stories of defeated tyrants, such as Pharaoh, while Ranjha invokes similar Hindu stories, such as that of epic hero Rama slaying the demon Ravana. Immediately, God grants their wish, and fire rages in the town until the frightened king sends troops to recover Heer from her husband's clan and restore her to Ranjha.

What is remarkable here is that even though the lovers are conventionally in the wrong because their love is adulterous, their love is vindicated by its steadfastness, so their curses have the same effect that the curse of an injured righteous person would have. Their curse sets the town on fire just as the faithful wife Kannagi's curse in the Tamil epic *Silappadikaram* sets the city on fire after the king unjustly executes her husband, thus separating her from him.

In both stories, the righteous protest is effective, but the punishment of the unjust is nevertheless followed by the death of the righteous victim. Kannagi dies, as do Heer and Ranjha. These deaths are represented as a protest against tyranny. Waris Shah's poem concludes with remarks such as: "Tyrants have ruled over vast domains,/shiftless they had to leave this world. . . . /evil everywhere is rife."[32]

"No One Can Separate Us"

At the same time, both texts also attribute the tragic events to destiny. The Goddess of Madurai explains to Kannagi that all that has happened is the result of previous births and was thus inevitable. Waris Shah comments: "This world is the play of children, Waris,/and dust at last must mingle with dust."[33] In both poems, the couples are reunited in death. Kannagi becomes a Goddess and meets her husband in heaven; Ranjha's soul joins Heer's in the upper air and both pass to life eternal.

In India, the idea that no one can separate those destined to be together is very powerful, and is interwoven with the idea of love-death. In the 1988 film, *Qayamat Se Qayamat Tak* (From One Apocalypse to Another) two feuding families react with great hostility when the daughter of one and the son of the other fall in love. The girl's family tries to forcibly marry her off to another man. The lovers elope and marry without witnesses, by garlanding one another in a ruined temple. This is a modern representation of a *gandharva* marriage. The woman's family sends hired killers to kill her lover but by mistake they kill her, whereupon he commits suicide. Her dying words to him, "Now nobody can separate me from you," invoke the normative Hindu idea that death does not sever attachments. Almost identical words are found in every same-sex couple's suicide note.

This idea of inseparability despite death is enshrined in a story about the paradigmatic good wife, Sati, consort of the destroyer God Shiva, who commits suicide to protest her father having insulted her husband. She is reborn as Parvati, and is determined to marry Shiva once more. The grieving Shiva, absorbed in asceticism, pays no attention to her, but she wins him by her intense austerities. Her suffering constitutes an ordeal that proves she is the only suitable consort for him, because she is the reborn Sati.

Among the Hindu Gods, Parvati and Shiva are, in many ways, normative spouses, who quarrel but always make up, and who prevent the balance of power from tipping either way. That their marriage is premised on death and rebirth is therefore significant.

Love is Strong as Death

The idea that love, like death, is an unstoppable force is found in many cultures. This idea is often found in conjunction with that of the equal force of hate. The Hebrew Song of Songs states it thus: "Love is strong as death; jealousy is cruel as the grave. . . . Many waters cannot quench love, neither can the floods drown it." (8: 6).

The idea that death may constitute a type of marriage is found in medieval Indian devotional traditions, both Hindu and Islamic. In these traditions, both marriage and death function as tropes for the devotee's mystical union with God. As discussed in chapter 3, both male and female *bhaktas* (Hindu devotees) imagine themselves as brides of God.

Death may consummate union with God, and thus be a kind of mystical marriage. A Muslim Sufi mystic's death is celebrated as his *urs* (Arabic for "marriage") to God, and at such a funeral, rejoicing, not mourning, is considered appropriate. At the tombs of major Sufis in South Asia, their death anniversaries even today are celebrated as wedding anniversaries. Several Hindu devotees are said to have died when they merged with the image of God in a mystic marriage: for example, Tamil woman poet Andal, Rajasthani woman poet Mira, and male Tamil poet Sambandhar.

Thus reports of lesbian weddings and joint suicides today play both on the Christian idea, well known in modern India, of "till death do us part," and Hindu and Sufi ideas of marriage as an eternal bond. In one newspaper report, a photograph of Leela and Urmila is captioned "Till death (and the police force) do us part."

Bury Us Together

Many couples, in their suicide notes, ask to be buried or cremated together. The bodies that could not be joined on the wedding bed are publicly united on the deathbed. Lalitha wrote in her suicide note: "I cannot part with Mallika. . . . Bury us together." In their joint suicide note written in blood, Mamata and Monalisa "expressed as their last wish to be cremated on the same pyre."[34] Similarly, Bina Wankhede and Nandita Gaekwad who tied their hands together and threw themselves in front of a moving train in Maharashtra, western India, asked that they be burnt on the same funeral pyre.[35]

Alan Bray, who has studied the marriage-like imagery on tombs of same-sex friends buried together in Western Europe from the fourteenth through the nineteenth centuries, concludes that the communities who built those tombs understood the relationships to be friendships, not sexual relationships. However, in the case of recent suicides in India, when families accede to the couples' wish for joint funerary rites, they know the nature of the relationship, because the couples make it clear in their suicide notes. When Sumathi and Geethalakshmi killed themselves, Sumathi wrote in a letter to her father, "We cannot live apart from each other." According to their last wish, their bodies were burnt on the same pyre at the Satyamangalam cremation ground, while family members wept aloud.[36]

Thus, when the cremation fire is lit for a same-sex couple, it is with the knowledge that they died in part because the wedding fire could not be lit for them. The funerary rite stands in for the wedding rite, and is itself a kind of wedding rite. Family and friends witness the rite, and so does the fire, which, in Hindu tradition, is a manifestation of the God Agni.

Among some Hindu communities at some times and places, widows burnt themselves or were burnt along with their dead husbands' bodies. The widow would dress in bridal finery, and the cremation would be celebrated like a second wedding, because the wife was accompanying her husband into death. Although idealized as a manifestation of conjugal love, this practice was more emblematic of the subjugation of women, as no parallel rite existed for men whose wives died before they did. The widow was known as a Sati or an embodiment of truth.

But a more egalitarian form of love and marriage in death was represented in South Asian love legends, such as those of Heer and Ranjha, Sasi and Pannu, Sohni and Mahiwal. In these legends, as in the Western legends of Romeo and Juliet or Tristan and Isolde, man and woman die for each other, and force society to confront the truth of unconventional love. Couple suicides today represent that more egalitarian form of love-death.

Seeking Rebirth

I quoted earlier the note left behind by two Nepalese women, Rekha and Gayatri, who wrote that they wanted to live as husband and wife "in the life beyond the grave." Hindu, Buddhist, and Jain traditions emphasize that humans should seek to escape the cycle of rebirth. However, much as love stories in Christian culture rewrote Christ's mandate that there is no marriage in heaven, Indic traditions of love also invert religious mandates.

Instead of longing for liberation from rebirth, lovers long to be reborn together. In a very popular Hindi film song, a lover sings, "So sweet, so intoxicating is our love that we will have to be born again and again."[37] Here, the conventionally distressing prospect of rebirth is re-envisioned as life after life spent happily together.

In contrast to Christian weddings, which unite spouses till death, Hindu weddings are understood to unite spouses for seven lifetimes, or even hundreds of lifetimes. Dying together acquires a special resonance in cultures, like the Hindu and the Buddhist, that believe in rebirth. Indian texts often attribute not just romantic love but any inexplicably strong attachment to a connection in a previous life.

Love suicide is a convention often represented in Japanese drama. In the famous play *The Love Suicides at Sonezaki* by Chikamatsu Monzaemon (1653–1725), Tokubei, aged 25, and his lover, Ohatsu, 19, a prostitute, exchange vows privately as husband and wife, and commit suicide (he kills her and then himself), praying to be reborn in paradise as husband and wife for ever. They also hear people discussing other love suicides.

The doctrine of karma, whereby actions in a former life shape the present life, is often misunderstood as causing fatalism. In fact, the doctrine is a dynamic one, since actions in the present life shape the next life. While one's present circumstances may be the result of actions in former lives, one's actions in this life can both reshape this life and have results for future lives. Karma and rebirth are thus invitations not to passivity but to action. Fidelity to an attachment thwarted in this life may result in its consummation in the next.

On the one hand, this theory explains the inexplicable chemistry between two people. On the other hand, it also holds out hope for the future. If two people are destined to be together because of an attachment in a former life, social opposition will not succeed in parting them. Taken to its furthest extreme, the theory suggests that if social hostility prevents them from living together, choosing to die together will reunite them in the next life.

At the personal level, the theory of rebirth functions to reassure individuals that their socially illicit feelings are justifiable. Hindu texts repeatedly insist that love is a force that cannot be contested: "Creatures are completely dependent upon connections in previous births, and this being the case, who can avoid a destiny that is fated to him, and who can prevent such a destiny befalling anybody?"[38]

Islamic Sufism too developed the idea that love is an innate human response to the divine, and therefore cannot be repressed. Nineteenth-century poet Ghalib famously expressed this through the metaphor of a self-igniting flame: "Love cannot be forced—it is a flame, says Ghalib,/Which cannot be ignited at will or quenched at will."

Another effect of such theories is to indicate that the lover has no choice and little control over love. The lovers' statements I quote throughout this book tend to use words expressing lack of choice or inability to control actions: "I *can't* live and sleep without you"; "We *cannot* live apart from each other"; "I loved her and I *couldn't* live without her" (emphases mine). But parents and other authorities are not always convinced by these claims. After Lalithambika and Mallika were saved from drowning, Mallika's elder brother was "reported to have said that the girls have agreed to try and forget each other."[39]

Rebirth Legitimizes the Illegitimate

More intriguing than parents who oppose same-sex marriage are those who come around to supporting it. Newspaper reports represent several parents participating in their daughters' weddings. In some cases, the weddings appear to have been elaborate affairs, attended by many guests. In none of these cases was any gay rights movement or organization involved. The arguments that convinced these parents were not, then, those that might have been put forward by gay rights advocates.

The family members quoted in newspaper reports represent themselves as wanting to make their daughters happy, and becoming convinced that they would be happy only if they married one another. Most of the reports, though, are brief. I would like to offer a hypothesis arising in part from a remark made by Urmila's neighbor, a woman called Sushila Bhawasar, who is a village schoolteacher, and who, along with her husband, also a school teacher, expressed indignation at the police authorities' suspension of the women from their jobs: "After all, what is marriage? It is a wedding of two souls. Where in the scriptures is it said that it has to be between a man and a woman?"

This remark is amazingly close in its philosophical assumptions to the reasoning of a Shaiva (devotee of Shiva) priest from India who performed the wedding of two Indian women in Seattle in 2002. He told me that when the women requested him to officiate at their wedding he thought about it and, though he realized that other priests in his lineage might disagree with him, he concluded, on the basis of Hindu scriptures, that, "Marriage is a union of spirits, and the spirit is not male or female."

Another, much better known, Hindu priest connected this argument of the ungendered spirit to the theory of rebirth. In her 1977 book, *The World of Homosexuals*, Shakuntala Devi recorded her interview with Srinivasa Raghavachariar, Sanskrit scholar and priest of the major Vaishnava (devotee of preserver God Vishnu) temple at Srirangam in South India. Sri Raghavachariar said that same-sex lovers must have been cross-sex lovers in a former life. The sex may change but the soul remains the same, hence the power of love impels these souls to seek one another.[40]

The fact that a similar explanation was given by both Shaiva and Vaishnava priests and by a female schoolteacher in a village indicates the overarching cultural importance of these notions of the spirit retaining its attachments but not its gender through various lifetimes. That these Indic notions help legitimize relationships that might otherwise be seen as illegitimate appears not just in these modern weddings but in some premodern texts, where parents come around to approving young people's socially inappropriate desire to marry.

Several such stories appear in the eleventh-century Sanskrit story cycle, the *Kathasaritsagara*. Some of these are ancient stories, found in earlier forms in Sanskrit drama from the first few centuries AD, and probably deriving from legends passed down in oral traditions. In the *Kathasaritsagara*, this set of stories begins with that of a prince who falls in love at first sight with a beautiful and brave Chandala girl, and wants to marry her. As the Chandalas are considered a very low caste, his love appears impossible. The prince's parents are perplexed when they see him pining away. His father, King Palaka, decides that the girl must not really be a Chandala and that "without doubt she was the beloved of my son in a former birth; and this is proved by his falling in love with her at first sight" (115).

What is remarkable here is not that the king turns out to be right. Fairy tales in many cultures conclude with the poor orphan turning out to be a royal scion in disguise. The specifically Indic element is the king's explanation for love at first sight, which proceeds thus: If the two were not suitable for each other, they would not have fallen in love. Since they have fallen in love, they must be suitable for one another. Therefore, they must "really" be of the same status.

This explanation calls into question the nature of "reality." The king's argument is based on the Hindu idea that attributes that appear real like caste and class (or even gender, as we have seen in chapter 3), may not be "real" because they change from one lifetime to another.

King Palaka tells his wife, "The minds of the good tell them by inclination or aversion what to do and what to avoid" (112). So, if his son is good, it follows that his inclinations are good too. Individual inclinations may sometimes be allowed to redefine social reality.

Marriage or Death

King Palaka tells his wife a story to prove his point. In this story, a Chandala (a caste considered very low) man and a princess fall in love when he rescues her from an elephant attack. The agonized man considers his love impossible and therefore unspeakable: "How can a crow and a female swan ever unite? The idea is so ridiculous that I cannot mention it or even consider it" (113). So he decides to kill himself. His logic is very similar to that of Anjali Thapa, the Nepalese woman mentioned earlier, who attempted suicide in December 2003, because she felt it was impossible to tell her parents about her love for a woman. The young Chandala lights a funeral pyre, but before entering it, prays to the God of fire, requesting that his sacrifice may result in his attaining the princess as his wife in the next life.

This idea connects to the one mentioned earlier—suffering proves one's love to be genuine, and results in attaining the beloved in an afterlife. The God of fire appears to the youth, and tells him that he is in fact his (the God's) illegitimate son who was adopted by a Chandala couple. The God also informs the princess's father of this in a dream, and the father then arranges his daughter's marriage to her lover.[41]

The narrator then tells another story of a fisherman and a princess who fall in love. The fisherman dies of anguish at separation, and the princess is determined to burn herself with his corpse. Her father then hears a divine voice telling him that the fisherman was a Brahman in a former birth and the princess his wife. The king agrees to the marriage, and the fisherman recovers his life because the princess gives him half her life.

Not all stories of illicit love end so happily. In another story told by the narrator, a merchant's daughter falls in love at first sight with a thief being led to execution, and declares that she will die with him. The text does not offer an explanation—it is simply assumed that her love springs from attachment in a former life. The merchant, anxious to save his daughter, offers a huge ransom to save the thief's life, but the king gets enraged with the merchant for making this request, and refuses it. So the thief is executed and the merchant's daughter enters the fire with his corpse.

Here we see two alternatives starkly outlined—marriage or death. If the parents, the government, and other authorities approve one's love, marriage ensues. If they persist in disapproval, true lovers would rather die than give in. In Western European drama and fiction, from the Renaissance to the nineteenth century, marriage and death were the two primary ways for a narrative to conclude. In Victorian novels, such as Thomas Hardy's *Jude the Obscure* (1894) and George Eliot's *Mill on the Floss* (1860), socially disapproved relationships that cannot be consummated in marriage culminate in death.

Several Indian films, from *Madhumati*, 1958, to *Milan* (Union), 1967, show rebirth uniting lovers parted by social forces. In the latter, a hugely popular Hindi remake of a Telugu film, newly weds on their honeymoon recover memory of their former life in which their love could not be consummated because of class difference and also because she was a widow. Social scandal led to their drowning together. But rebirth reunites them.

Love suicides have been an institution in Japan since the late seventeenth century. According to the Lotus Sutra, lovers who die together are reborn together in paradise.[42] A Japanese Christian couple whose parents refused them permission to marry, drank poison together in the 1950s, leaving a note that said, "We go to heaven together, to marry in the presence of God."[43]

Dreaming Up Social Change

The role played by dreams in many of these stories is telling. Parents, anxious to save their children's lives, dream that those children's illicit desires are in fact licit. Ashis Nandy mentions that many of India's folk heroes have "convenient dreams" that "sanction participation" in novel enterprises that imply "defiance of conventional authority."[44]

I would add that dreams also allow individuals, besieged by society, to turn inward and dwell on significance that extends beyond outward appearance; dreams thus enable a re-envisioning of the future. Anxious parents' dreams reveal that the illicit lover was the child's spouse in a former birth and was also differently categorized, that is, belonged to a different caste and class. This enables the parents to recategorize him or her in the present life. This demonstrates, as is clear from many other narratives as well, that caste, like gender, changes from one life to another and that the spirit has no caste just as it has no gender.

What is most interesting, though, is that once the parents realize the children are ready to die rather than give up their lovers, they decide to overlook the indisputably low caste and class of the lover. This accommodation is similar to that made by many parents of female couples discussed in this book, who decide to accept a woman as their daughter's groom.

Equally interesting is the way other social authorities react. Parents may be emotionally inclined to make compromises for their children's happiness, but can they persuade others to accept these compromises? We saw that in one story, the king refuses the distraught father's request to spare his daughter's beloved's life. In the frame story, King Palaka comes up against similarly obdurate authority figures.

Palaka asks the Chandala girl's father to give her in marriage to the prince. The father says he will give his daughter only to the man who makes 18,000 Brahmans of the city eat in his house, that is, he insists on social acceptance. The king summons all the Brahmans in the city, tells them the story, and orders them to eat in the Chandala's house. The Brahmans are not willing to pollute themselves by obeying his order. So they go to the shrine of Shiva and perform self-torture there, in protest. The king has to choose between his son's death and the Brahmans' death.

The story thus sets up an interesting situation. The king, influenced by paternal feelings, agrees to a cross-caste marriage. But the religious authorities are not willing to sanction it. They, however, are also caught in a dilemma. Since they depend on the king's patronage for their livelihood, they cannot afford to annoy him. In this respect, the situation is different from that in the earlier story, where the king, who is superior to the merchant, can afford to refuse the merchant's request. The Brahmans' dilemma is resolved when they have a dream—Shiva tells them to eat in the Chandala's house because the Chandala was formerly a heavenly being (Vidyadhara).

The Brahmans agree but do not altogether renounce their rules of pollution and purity; they modify them for the occasion. They tell the king to tell the Chandala to get pure food cooked for them outside the Chandala neighborhood. The king builds a new house for the Chandala, and pure cooks (presumably Brahmans) cook the food there. The king thus goes to great expense to legitimize his son's marriage.

We see here a series of delicate social negotiations, undertaken to legitimize disapproved marriages. These negotiations depend on the shared philosophical premise that love arises from attachment in former births and is therefore irrepressible. Despite this shared premise, the negotiations do not always succeed. Parents are most susceptible to such legitimization when their children's lives are at stake; other social figures may give in if they are dependent on the parents. Much depends on how powerful the parents are, and how much effort they are willing to make on their child's behalf. Breakdown of negotiations may result in the lovers' death; their success results in the disapproved marriage being approved.

Chapter 5

A Second Self: Traditions of Romantic Friendship

He who is not touched by love walks in darkness
—Pausanias, in Plato's Symposium

He was thinking of the irony of friendship—so strong it is, and so fragile. . . . Abram and Sarai were sorrowful, yet their seed became as sand of the sea . . . But a few verses of poetry is all that survives of David and Jonathan. . . . man is so made that he cannot remember long without a symbol; he wished there was a society, a kind of friendship office, where the marriage of true friends could be registered.

—E.M. Forster

In 2001, a Hindu priest in the Srivaishnava lineage conducted a friendship ceremony for two Hindu women in Sydney, Australia. A gay man wrote to ask him if a same-sex couple could have a *gandharva* marriage. The priest replied, "Marriage (*vivaha*) by definition is between male and female, the purpose being reproduction and the performance of one's duties as householders. There is a commitment ceremony for friendship as described in the *Ramayana* between Rama and Sugriva—it is not the same as a 'marriage' but has some of the same ritual elements—holding hands, exchanging garlands and walking around the sacred fire—taking seven steps together etc., the purpose being to confirm and validate one's commitment to the friendship-relationship. So the question of *gandharva* or any other form of 'marriage' cannot arise within the Hindu context between members of the same sex."[1]

As we have seen, several Hindu priests disagree with him, and do perform wedding ceremonies for same-sex couples. But the fact that the friendship ceremony and wedding ceremony share many ritual elements points to the overlapping and inextricable nature of marriage and friendship. Because same-sex friendship in traditional societies has been idealized more than cross-sex friendship, same-sex marriage today, even more than cross-sex marriage, is a place where ideas about marriage and ideas about friendship meet. Those who debate same-sex marriage today are frequently unaware of the complex histories of friendship, and its influence on ideals of love and marriage.

The Swayamvara Friend

The word "friend" in English has lost the intense charge it held until the nineteenth century. In medieval and Renaissance Europe, a lady's lover was termed her "friend." A parallel in Urdu/Hindi is the word "*yaar*" which used to mean a lover of either sex, but now is used more as "buddy" is in American English, although it still retains its earlier meaning as well.[2] The word is derived from the Persian *yar*, and also from the Sanskrit *jara* (woman's male lover).

When writers like Jane Austen in the eighteenth and early nineteenth centuries refer to a woman's "friends," they include her family in that category. The situation is now reversed in the West, with biological family being separated from and treated as more important than nonbiological friends. Paradoxically, it is in modern culture that biology trumps choice.

In medieval Sanskrit texts, a special friend is termed a *swayamvara* or "self-chosen" friend, different from run-of-the-mill friends. Aristotle also distinguished the one "true friend" from many others with whom one may be "friendly." The word *swayam* means "self" and *vara* means "boon," "wish," or "desire," and also comes to mean "bridegroom" or the "desired one."

The use of *swayamvara* as an epithet for a special friend indicates the overlap between marriage and friendship. In the eleventh-century *Kathasaritsagara*, this term is used for both a male's male friend (*sakha*) and a female's female friend (*sakhi*). Such a friendship is referred to as "*janamantara*" (continuing from birth to birth); like marriage, it is based on reciprocity, selfless devotion, and sacrifice; as in ideal marriage, the partners live and die together. When a heavenly female, Somaprabha, sees princess Kalingasena and is spontaneously attracted to her, she decides that they must have been linked in a former birth and that she should therefore choose Kalingasena as her *swayamvara sakhi* (self-chosen friend). A number of words for love are used to indicate their feelings—*prema*, *sneha*, *priti*.

In TV sitcoms in the United States today, male friends' hesitation to express their feelings for each other, and their reluctance to use the word "love" becomes the source of much humor. This is because exclusivity and intensity are today seen as hallmarks of heterosexual love, not friendship, especially not male-male friendship, which, in the United States today, is fraught with anxiety about homosexuality. Pre-twentieth century texts did not assume that only spouses have a monopoly on love. They recognized that love and friendship go together—Jane Austen named her first novel, "Love and Friendship," and all her novels revolve around these two as inseparable experiences. As Roman statesman Cicero (106–43 BB) points out in his very influential Latin dialog on friendship: "Both words, *amor* [love] and *amicitia* [friendship], come from *amare*, to love. And love is precisely the nature of the affection you feel for your friend."[3]

Some of the most famous English encomia of love and marriage, now appropriated by heterosexual marriage, were originally written in the context of same-sex relationships. "Let me not to the marriage of true minds/ Admit impediments. Love is not love/Which alters when it alteration finds," and "Shall I compare thee to a summer's day?/Thou art more lovely and more temperate" are among such famous declarations; both occur in Shakespeare's sonnets to a beautiful young man, Mr. W.H. As late as the nineteenth century, Alfred Tennyson, writing of his deceased friend Arthur Hallam, made another such famous statement: " 'tis better to have loved and lost/Than never to have loved at all."

In the modern era, the cultural ideal of union shifts from friendship to heterosexual coupledom. For example, the United States has had unmarried Presidents in the past, but today it would be nearly impossible for a single man to be elected President. Heterosexual coupledom is seen as central to the good and happy life, which the President is supposed to model for the people. In India, where the premodern still permeates the modern, there is no such requirement. Presidents' and Prime Ministers' wives have not acquired the cachet that American "first ladies" have. India had a widow, Indira Gandhi, as Prime Minister, and, as recently as 2003, a lifelong bachelor, Vajpayee, as well. Furthermore, friendship is still idealized both in private and public spheres, as seen in the case of Jayalalithaa, chief minister of Tamil Nadu, whose friendship with another woman constitutes part of her public persona.

"Mother" and "Small Mother"

After Jayalalithaa, a single woman, became chief minister of Tamil Nadu in south India in 1991, her close friend, Sasikala Natarajan, a married woman who ran a video supply store, moved into the Chief Minister's official residence. The two friends are respectfully addressed as "*Amma*" (Mother) and "*Chinamma*" (Small Mother).

In the English language media, Sasikala is described as Jayalalithaa's "aide," "close friend," and "live-in confidante."[4] The Tamil media terms her Jaya's *thozhi*. *Thozhi* is the equivalent of the Sanskrit *sakhi* or female friend. In premodern narrative, royal and aristocratic heroines generally have a *thozhi*.[5]

In July 2001, after her reelection as Chief Minister, Jayalalithaa, accompanied by Sasikala, went to the Sri Krishna temple at Guruvayoor in Kerala, to offer an elephant in fulfillment of a vow. On another occasion, they visited the Shiva temple at Mylapore, Chennai. It is the custom that after the temple tank is purified, a newly wed couple takes the first dip in the tank. On this occasion, Jayalalithaa and Sasikala took the first dip and poured water on one another's heads, in the presence of thousands. In March 1992, they took a bath in the tank at the Kumbakonam temple, Tamil Nadu, on the occasion of the major festival held there once every twelve years, which coincided with Jaya's birthday that year. Sasikala poured water from 21 silver vessels on Jaya's head and she in turn poured water on Sasikala.[6] The national media reported all these events, with photographs.

Jayalalithaa's party-conferred title, "*Puratchi Thalaivi*"(revolutionary leader) enshrines her relationship with party founder M.G. Ramachandran whose title was "*Puratchi Thalaivar.*" However, in recent years, she has often been represented in partnership with Sasikala. Aside from Jayalalithaa's signature garment, a cape, both women dress similarly, in silk saris and gold jewelry. In many photographs, they stand side by side, wearing matching garlands. Jayalalithaa and Sasikala's is a cross-caste relationship. Jayalalithaa is a Brahman and Sasikala a Thevar. Referring to Sasikala's influence, the AIADMK's opponents sometimes dub it a Thevar party.

There have been few frontal attacks on the relationship although there have been snide remarks. For example, after Jayalalithaa entered into a coalition with the BJP, she was accused of twisting Prime Minister Vajpayee's arm. The online news magazine, *Rediff*, commented, "Mother twisted Vajpayee's arm indeed! The Wholly Mother, good sir, is not interested in even holding anyone's hand (except Sasikala's), let alone twisting it!"[7]

In online chat rooms and web pages, she has been viciously attacked. When a woman, defending Jayalalithaa, claimed that male chauvinists always call single women names, a male correspondent replied: "First of all Jayalalitha is not a woman. That is why Sasikala is always with her (him)."[8] Jayalalithaa, however, continues to survive such attacks, and to project her friendship as a positive aspect of her image.

Gender and Politics

Jayalalithaa's mass following includes millions of women. In 1991 the party brought 26 women into the Tamil Nadu assembly, the largest number ever elected. Jayalalithaa introduced a 30 percent quota for women in all police jobs, and set up 57 all-women police stations, as well as all-women libraries, stores, banks, and cooperative societies.

In 1996 she lost the elections, largely because of the corruption of which she and Sasikala were accused. While several Indian politicians have outdone Jayalalithaa in corruption, none has suffered the opprobrium that she has. Her gender clearly figures in the excessive virulence with which the national media portrays her. The prime symbol of her corruption became the ostentatious wedding of Sasikala's nephew, to which 400,000 people were invited. Jayalalithaa had adopted this young man, and arranged his marriage. Both women were arrested on charges of bribery and appropriating public funds.

The media decided that Jayalalithaa's career was over. But in 2001 she was again elected Chief Minister.[9] On the one hand, her gender seems to work for her, with her followers celebrating her both as a Goddess figure—she has been portrayed as Shakti, the Virgin Mary, and Bharat Mata—and as a wronged victim fighting for justice, like Draupadi, heroine of the *Mahabharata*. On the other hand, gender also works against her, with the intelligentsia stereotyping her as an Imelda Marcos and a Margaret Thatcher.

In 1996, Jayalalithaa conducted the marriage of 5004 couples, acting as a "non-religious high priestess" and gave *talis* to the brides.[10] Each couple also got silk clothing and Rs.10, 000. Though she presides over weddings, Jayalalithaa insists on women's autonomy: "It is my firm conviction that a woman should marry only if she wants to raise a family, not simply because she needs a man to support her."[11]

Same-Sex Union and Political Alliance

How has Jayalalithaa survived in politics without sacrificing her friendship? One reason is that commitment to friendship is a central value in Indian culture. The ability to maintain a loyal friendship, regardless of one's marital status, is admired as indicating an integrated personality.

Some theorists claim that male bonding in premodern societies is really about political alliance and male domination.[12] While agreeing that male bonding may reinforce male domination, I argue that this dimension does not erase other dimensions of union, such as love and friendship. Nor is same-sex bonding exclusive to men.

In the ancient Sanskrit epic, the *Ramayana*, when King Rama and monkey king Sugriva become friends, they ratify their commitment with a sacred ritual. Monkey warrior Hanuman lights a fire. Rama and Sugriva exchange vows of friendship and walk around the fire. The fire God thus becomes a witness to their vows.

This friendship ritual is very similar to the central rite in most Hindu weddings, which consists of bride and groom walking around fire. The episode of Rama and Sugriva walking around fire represents an ideal. The normative political alliance was ideally also a personal bond and a sacred commitment. Just as the devotion of a political leader to his or her spouse may be valued as normative, so may his or her devotion in friendship. Jayalalithaa's loyalty in friendship (to MGR and Sasikala) adds to her luster in the eyes of her followers.

Same-Sex Union and the Good Life

Both in Indian and Western thought, friendship has long been seen as an essential, even *the* most essential, component of the good life. In the *Nicomachean Ethics* Aristotle claims: "without friendship no one would choose to live, though he had all other goods"; later, many Christian thinkers echo this claim.[13]

Ideals of friendship, both in Indian and Western thought, are very similar to ideals of marriage. Friendship, in both premodern Europe and India, is defined as a long-term, exclusive, intimate, non-biological relationship, which entails friends living together and sharing everything. Friends exchange vows of eternal fidelity. They give without being asked, and make sacrifices for one another. Friends sometimes die together; if one dies, the other feels incomplete. Friends are often buried or cremated together, and others commemorate them together. All these are features also ascribed to an ideal marriage and an ideal love relationship. Saleem Kidwai has analyzed the centrality of male-male friendship and love not only in the spiritual economy of Sufism but also in the urban economies and cultures of the medieval Indian Islamicate.[14]

An obvious difference between marriage and same-sex relationships is that marriage generally leads to procreation. In Plato's *Symposium*, Socrates' teacher Diotima argues that same-sex relationships are also reproductive, because same-sex lovers jointly create ideas, books, and institutions, which make them immortal, just as children make parents immortal, and which also live longer than regular children and benefit humanity more: "Who, when he thinks of Homer and Hesiod and other great poets, would not rather have their children than any ordinary human ones?"[15]

Living Together

Aristotle states: "there is nothing so characteristic of friends as living together."[16] For this reason, he says, one should not try to have as many friends as possible but only "as many as are enough for the purpose of living together."[17]

Aristotle's idea of friendship persisted in Western thought until the modern era. In Cicero's dialog on friendship, Laelius says of his dead friend Scipio, "I feel convinced my life has been a good one—because I have spent it with Scipio" and "We shared the same home, we ate the same meals, and we ate them side by side."[18]

The ideal carried over into Christian thought. St. Augustine, who had a passionate relationship with his mistress, nevertheless describes his friendship with his male friend as "sweeter to me than all sweetnesses that in this life I had ever known."[19] After his conversion, Augustine broke his engagement to marry a woman, and, instead, lived with a small group of close friends, along with his mother and his son.

Many premodern Indian texts show friends spending their lives together. In the *Panchatantra*, an ancient Indian text that traveled to Europe, with versions appearing in Greek, Latin, German, Spanish, French, English, Armenian, Hebrew, and Slavonic languages between the eleventh and eighteenth centuries, friendship is represented as the basis of good government, and also of the good life. Translator Chandra Rajan remarks that in this text, "it is friendship that is given a special place and set above all other relationships."[20] In this text, four male friends, a mole, a crow, a deer, and a tortoise, live together and make sacrifices for one another.

In the *Kathasaritsagara*, when Vasudatta, a merchant's son and Pulindaka, a tribal chieftain, choose one another as *swayamvara* friends, Pulindaka leaves his forest home and lives in Vasudatta's home. Vasudatta repeatedly says his life is happy because of his wife and friend, and, later, his son. Happy life is thus based on both marital love and the love of a friend. Whether or not Pulindaka marries is not mentioned. His happiness appears to be based entirely on this same-sex friendship.

Today, in the West, friends live together only as a premarital phase. TV shows like *Friends* and *Seinfeld* nostalgically depict adult friends continuing to live together.

Friend as Second Self

Aristotle notes that some people think we choose as friends those who are like us, while others think we choose those who are different from us. The tension between these two ideas of attraction persists in discussions of love—do opposites attract or does like attract like? Aristotle insists that we gravitate toward that which is like ourselves, therefore the good will seek the good and the bad the bad, but true friendship can exist only between good persons.

Aristotle, unlike Plato, sees women as inherently inferior to men, so, according to him, marriage is a union of dissimilars and friendship a union of similars. Aristotle argues that marriage, like the parent-child relationship, is friendship between unequals, while friends, even if socially unequal, are equalized by friendship. This view was highly influential in premodern Europe.

Plato (who, in my view, makes no significant difference between friends and lovers, seeing the latter ideally developing into the former over time) makes a somewhat different claim. In his *Symposium*, while most speakers emphasize that like is drawn to like, and that true love and friendship subsist between the virtuous, Socrates insists that we are drawn to that which we lack. Some of Plato's speakers indicate that friendship and love can exist between women too.

Premodern Indian texts stress likeness as the basis for friendship, love, and marriage. While some texts uphold the norm of likeness in external attributes such as wealth and beauty, many others stress a deeper likeness that transcends external disparities such as caste, and harks back to a former birth. Even though women are not men's social equals, love stories often represent them as spiritual equals.

A modern Jain marriage manual opens with a verse stating that friendship and marriage are appropriate between similars.[21] The beauty of paradigmatic male friends Krishna and Arjuna (who belong to different castes) is frequently mentioned in the *Mahabharata*: "like a couple of risen suns . . . like the bright and many-rayed moon and the sun risen after dispelling a gloom" (*Karna Parva*, XCIV). Indian texts stress likeness in beauty between same-sex friends as well as cross-sex lovers, while ancient Greek texts ascribe beauty primarily to the younger male beloved, not to the older male lover.

Divine friends, like divine consorts, are commemorated together and worshipped together. The *Gita* ends with the statement that victory resides wherever Krishna and Arjuna reside. Similarly, Lord Ayyappa, divine son of the male Gods Vishnu and Shiva, has an inseparable friend Vavar, who is a Muslim. Ayyappa is said to have asked his foster father to build a shrine for Vavar, saying, "Consider Vavar as myself." Today, Ayyappa's temple in Kerala, the focus of a huge annual all-male pilgrimage, has a shrine for Vavar. All pilgrims, Hindu and Muslim, are supposed to offer pepper to Vavar.[22] Friendship thus crosses barriers of caste, class, and religion.

Equality and Sharing

This stress on likeness is an important example of the way ideals of same-sex love have influenced later ideals of heterosexual marriage. Today's glorification of likeness and equality (rather than difference) between spouses borrows from the ancient tradition of celebrating likeness between same-sex friends. For example, today a wife is called her husband's "better half," but the phrase is not applied to friends. Before the twentieth century, it was commonplace to think of one's friend as one's other half or a second self. Thus, Cicero writes, "When a man thinks of a true friend, he is looking at himself in the mirror."[23]

If the friend is a second self, do all one's worldly goods belong to him/her? Most philosophers of friendship would answer in the affirmative. Sixteenth-century French philosopher, Montaigne, citing Aristotle, defines true friends as "one soul in two bodies."[24] Emphasizing that this kind of friendship is rare and not to be confused with everyday friendliness, Montaigne writes that true friends have "everything . . . in common between them—wills, thoughts, judgments, goods, wives, children, honor, and life."[25] Indian narratives too dwell on friends' inseparability and sharing of possessions. As discussed in chapter 1, Montaigne argues that the ideal of shared matrimonial property is modeled on the ideal of same-sex friends sharing everything.

The ideal of friendship equalizing social unequals is found in the story of Muslim king Mahmud and his beloved slave Ayaz. Many medieval Indian Sufi poets celebrate their love as a symbol of the perfect love that crosses the gulf between God and devotees.[26]

A parallel in Hinduism is the story of Krishna and Sudama, childhood friends. While Krishna goes to Mathura and becomes king of the Yadavs, Sudama, a Brahman, lives in dire poverty in the village. Sudama's wife advises him to ask Krishna for help, so he reluctantly sets out, carrying a present of parched gram. When he arrives at the palace gate, Krishna runs out to embrace him. He insists on eating the gram. Sudama does not ask for help, but when he returns to his village, he finds his hut replaced with a mansion filled with wealth. The story shows how friendship crosses class and caste boundaries, and also indicates how God's love for humans raises them to his level.

The Female Friend

Indic narrative typically depicts friends as social unequals. The heroine's female friend (*sakhi* in Sanskrit) and the hero's male friend have an important place in love literature, both narrative and lyrical. They are found in fifth century Tamil Sangam love poetry, as well as in ancient Sanskrit poetry. These friends are often social subordinates—courtiers or servants. The *sakhi* in particular remains a stable figure through the huge corpus of medieval mystical love poetry, both canonical and folk, in most major Indian languages. Paintings, sculpture, and modern cinema also give the *sakhi* a prominent place as confidante and go-between.

In the story of divine lovers Radha and Krishna, Radha, although not Krishna's wife, functions as his consort, and her women friends are sometimes his lovers and thus her subordinate co-consorts. Some friends are her handmaids who love her as much as or more than they love Krishna. In some traditions, devotees, male and female, identify with Radha's women friends, as a gesture of humility.

In miniature paintings that derive from the Radha-Krishna tradition, the *sakhi* is often present not only when the heroine is alone, pining for her male lover, but also when the hero and heroine engage in sexual play. Indian lovers, unlike European lovers, are often depicted not in solitude but in a crowd. This tradition continues in modern cinema, where hero and heroine sing love duets surrounded by friends.

In late medieval miniature paintings, the space occupied by the *sakhi* often becomes eroticized. She frequently helps the heroine to bathe, dress and undress, and drinks or plays games with her; sometimes, they engage in mutual sexual play. Groups of girlfriends hunt, fly pigeons, dance, or bathe together, and their eyes meet erotically while one touches the other's breasts or embraces her. The common motif of the *sakhi* holding up a mirror to the heroine symbolizes their symbiotic relationship. She can be read as an aspect of the self, close and accessible, in contrast to the distant divine male lover. The heroine's dialogue with her is like a dialogue with oneself.

Except where the women's interaction is explicitly sexual, modern commentators almost always title these paintings and poems in a way that downplays the significance of the interaction. In eighteenth-century Hindi *Reeti* poetry, the girlfriend often expresses ardent admiration of the heroine. She makes comments like: "Heavens!/ How much beauty has god given her!/ Even I am bewitched by it, dear lad,/how much more/you!"[27] The translator has here inserted the words "how much more you." Literally, the original says simply: "Looking at that unique girl, I am entranced. How much sweetness God has given to her beautiful form."

As discussed in chapter 8, the important move Urdu Rekhti poetry makes is to explicitly sexualize the *sakhi*, a figure it inherits not from the Perso-Urdu ghazal but from medieval Indic language love poetry.

Spontaneous Attraction is the Hallmark

In premodern texts, friendship is more like romantic love than like marriage in one respect—one's family generally chooses one's marriage partner while one chooses a friend or lover for oneself. The family looks for social and economic compatibility when choosing a spouse; the individual looks for emotional compatibility when choosing a friend or lover.

In both European and Indian traditions, the experience of being drawn to a friend or lover is represented as miraculous. Plato, in the *Phaedrus*, describes the male lover and male beloved overwhelmed with wonder as they are drawn together. The role played by chance in any such meeting is inscribed as miraculous. In Indian medieval romances, both Hindu and Islamic, when two people (of whatever sex) are drawn together, each wonders if the other is a heavenly being. Often, they are attracted even before they meet, from reports they hear or pictures of each other that they see. In Hindu texts, this wonder is resolved in the conviction that they were connected in a former birth.

In Christian contexts, this kind of relationship is seen as predestined by heaven. Montaigne declares of himself and his friend La Boetie, "We sought each other before we met because of the reports we heard of each other, which had more effect on our affection than such reports would reasonably have; I think it was by some ordinance from heaven. . . . At our first meeting, which by chance came at a great feast and gathering in the city, we found ourselves so taken with each other, so well acquainted, so bound together, that from that time on nothing was so close to us as each other."[28]

Even though Montaigne carefully distinguishes his friendship from ancient Greek male-male love, his famous explanation of his love for La Boetie can only be called romantic: "If you press me to tell why I loved him, I feel that this cannot be expressed, except by answering: Because it was he, because it was I."[29]

In the *Kathasaritsagara*, descriptions of two men's or two women's attraction to one another are very similar to descriptions of a man and woman's attraction. The explanation for both is the same—they were attached in a former birth. Bandits capture the merchant Vasudatta and are about to sacrifice him to their Goddess, but bandit chief Pulindaka feels immense pity and affection for him the moment he sees him. The narrator comments: "Affection [that arises] in the heart without a cause speaks of love [persisting] from a former birth."[30] Vasudatta has the same feelings when he first sees his wife. They later come to know that in a previous life, she was his wife and Pulindaka was his friend.

In the same text, the description of two women's spontaneous attraction is even closer to conventional descriptions of men and women falling in love. Both types of narrative emphasize how physical beauty causes love. Heavenly female Somaprabha is flying through the sky when she sees princess Kalingasena playing on her palace roof. Kalingasena is beautiful enough to attract even an ascetic. Love is born in Somaprabha's heart. She wonders if Kalingasena is the Goddess of love or a heavenly nymph. She decides that since her heart is agitated by love for Kalingasena they must have been friends in a former life. She approaches Kalingasena who in turn appreciates Somaprabha's wonderful beauty and embraces her.

Intimacy—Physical and Emotional

In some ancient Greek texts, male-male love and friendship shade into one another, and there seems to be no great anxiety to distinguish them. Thus, in Plato's *Symposium*, it is mentioned in passing that Pausanias and Agathon are lovers, but the discussion of love never gets bogged down in a discussion of sex. When Alcibiades offers himself sexually to Socrates, Socrates neither accepts nor absolutely refuses the offer. He merely says, "at some other time then we will consider and act as seems best about this."[31] The deferral demonstrates not only his temperance and indifference to sensual pleasure but also the relative lack of anxiety attached to the question of whether or not to have sex.

In later Christian texts about romantic friendship, more anxiety is apparent, which indicates the impending historical displacement of same-sex love by cross-sex marriage. Thus, Montaigne at first argues that friendship is superior to sexual love precisely because it is not fleshly. He then admits that if bodies could be involved in friendly love as well as souls, "it is certain that the resulting friendship would be fuller and more complete."[32] As soon as he mentions bodies, the anxiety of gender appears. Since he agrees with ancient authors that women are not capable of exalted friendship, the only way to involve both bodies and souls would be a sexual relationship with a male friend. He immediately declares: "And that other, licentious Greek love is justly abhorred by our morality."[33]

Too much of a humanist to simply cite scripture as a reason, he instead casts about for a logical reason. He contrasts the inequality of age and beauty between ancient Greek male lover and male beloved, with the equality of friendship. Unlike some of his modern followers, who stop there, he is enough of a classicist to know that this inequality in Greek texts was not supposed to be permanent nor was it always apparent.

So Montaigne goes on to say that the ancient Greek ideal describes male-male love beginning in sexual passion but culminating in the equality of friendship. In acknowledging this, he comes close to undoing his initial argument. By his account, the only difference between his ideal of friendship and the ancient Athenian ideal is that his excludes growth and change but, by his own admission, also excludes the greater completeness made possible by an engagement not only of souls but also of bodies.

The ideal of marriage developing in Montaigne's time was beginning to make precisely that claim of completeness for heterosexual marriage. Edmund Tilney, in his 1568 dialog, *The Flower of Friendship*, sees marriage as the highest form of friendship because it involves companionship of bodies as well as souls. The modern Western ideal of marriage, like the ancient Athenian ideal of same-sex love, incorporates the developmental model, with youthful sexual passion growing into mature friendship.

In many premodern texts, a man and woman's sexual union is signaled by their marriage, sharing of a bed, and the birth of children. Explicit details of sexual intercourse are not usually part of romantic narrative. Embraces and kisses are more likely to be described than genital activity. Similarly, when we are told in the *Symposium* that two young men, Pausanias and Agathon, are "of the manly type," and are lovers, no further details of their sexual activities are given or expected. When close friends are described as physically intimate or are compared to lovers, that is not the equivalent of terming them lovers, but neither does it rule out the possibility of their being lovers.

In the *Panchatantra* beast fable of four close friends, when the crow embraces his friend the tortoise, the narrator comments:

> "Can sandal-paste blended with chill camphor
> or snowflakes delightfully cool, compare
> with the refreshing touch of a friend's body?
> They are not a sixteenth part of this delight." (II: 45)[34]

The metaphor of sandalwood is often used in Indian texts in the context of erotic contact, for example, princess Sasiprabha, in the *Kathasaritsagara*, narrates how she felt when her lover Manahsvamin carried her in his arms: "Then contact with his body made me feel as if I were anointed with sandalwood ointment, and bedewed with ambrosia . . ." (II: 301–07, Vetala 15). After embracing, the crow and mole sit together, "their bodies still thrilling with happiness" (*pulkita shariro*).[35] The narrator remarks that the embrace of a friend after many days is invaluable—no price can be set on it.

Dying Together: One Soul in Two Bodies

Both great love stories and great friendship stories often end with the partners dying together. For example, after a long and happy life together, Pulindaka and Vasudatta grow old and retire to the forest. Vasudatta remembers his life in an earlier incarnation, and throws himself from a cliff, meditating on Shiva. His dying wish is that he may have the same wife and friend in his next birth. His wife and friend jump off the cliff too, and die with him. Thus the friend's devotion exactly parallels the wife's.

In the ancient Greek myth of Orestes and Pylades, known to most schoolboys in premodern Europe, each tries to die in place of the other. The names of these two became proverbial as representing perfect friendship. In the *Symposium*, the names of male couples who died for each other, like Achilles and Patroclus, Harmodios and Aristogeiton, are linked with that of Alcestis who gave up her life to save her husband from death. The Theban band, made up of male couples, lover and beloved fighting side by side, was famous as an elite military unit; a speaker in the *Symposium* remarks of it, "An army of lovers cannot fail."

Death often functions as a divine test of friendship. In a famous medieval European story, Amis saves his friend Amile's life. Later, Amile, at an angel's command, kills his children to cure Amis of leprosy. This is a test, and the children are miraculously restored to life. Similarly, in the *Kathasaritsagara*, when bandits capture merchant Vasudatta to sacrifice him to the Goddess, bandit chief Pulindaka decides to sacrifice himself in his place. The Goddess is pleased, and stops him, also offering him a boon. Pulindaka asks that he and Vasudatta may be friends in the next life too.

Paradoxically, the ultimate sacrifice may be to live on after one's friend dies. When Shakespeare's Hamlet is dying, his friend Horatio is about to drink poison. Hamlet stops him, asking him to vindicate Hamlet's reputation by telling the true story. The connection of a friend to one's story is differently figured in Valmiki's *Ramayana*, where, when Rama is about to go to heaven, all his friends, including the monkeys, follow him, but Rama tells his dearest friend Hanuman to remain alive as long as Rama's story is told on earth. That is why Hindus do not kill monkeys, considered embodiments of Hanuman.

St. Augustine, describing his anguish when his friend died, recalls Orestes and Pylades, "who both wanted to die together, each for each, at the same time, since not to live together was worse to them than death."[36] But St. Augustine also feels that his friend lives on in him: "For I felt that my soul and my friend's had been one soul in two bodies, and that was why I had a horror of living, because I did not want to live as a half-being, and perhaps too that was why I feared to die, because I did not want him, whom I had loved so much, to die wholly and completely."[37]

Similarly, Montaigne says that the whole of his life, though reasonably happy, seems to him "nothing but smoke, nothing but dark and dreary night," compared to the four years he enjoyed "the sweet company and society" of La Boetie. Quoting several classical poets' similar feelings, he declares that since he lost his friend, he only "drag[s] on a weary life."[38]

Buried Together

As discussed in the previous chapter, several female couples that committed or attempted suicide together in the last two decades asked to be buried or cremated together. Through this wish, they modeled their deaths on the great love stories of the past, both cross-sex and same-sex.

Alan Bray has documented how, from the fourteenth to the nineteenth centuries in Europe, many same-sex friends made arrangements to be buried in the same monument or even the same tomb, a practice normally reserved for family members. He points out that for Christians, this means that at the Resurrection, friends will be reunited as they emerge from the grave. When Amis and Amile finally die together in battle, they are to be buried in two specially constructed churches, but, before the funeral, Amile's body in its coffin is found lying next to Amis's. The narrator exclaims: "Behold then this wondrous amity, which by death could not be dissevered!"[39]

Sufi poet Jamali (died 1535/6) lies in a tomb in south Delhi. Next to his grave is another, where, according to tradition, his beloved disciple Kamali is buried. The tombs are popularly known as Jamali-Kamali—the similarity of names is similar to that of Amis and Amile. So also, Punjabi Sufi poet Shah Hussayn (1539–1599) is buried in Lahore, and his beloved companion, Madho Lal, who died many years later, is buried next to him. The anniversary of Hussayn's death is celebrated annually at their tombs.

Rituals of Union: Vows and Witnesses

Both opponents of gay marriage and radical gay theorists criticize gay weddings for mimicking straight weddings. What they forget is that there is an equally long tradition of publicly ritualizing same-sex unions, whether these are characterized as marriage, friendship, or some other type of union. These rituals are not specific to either cross-sex or same-sex union—they include exchanging gifts and vows, eating together, holding hands, and invoking witnesses, human and divine.

Exchanging vows and joining hands are part of wedding ceremonies the world over. In many Hindu ceremonies, the vows take the form of conditions imposed by each spouse on the other. For example, the groom tells the bride to be hospitable to his friends, and she tells him not to keep secrets from her or reveal her secrets to others.

Indic friendship narratives depict same-sex friends exchanging vows and joining hands. Princess Kalingasena and her friend Somaprabha, after they are attracted to each other, embrace, converse, and then swear friendship with joined hands. In the sixteenth-century Indian Sufi romance, *Madhumalati*, two women become ritual sisters by embracing and exchanging vows.[40] A ritual of making a person of the same sex one's dharma-*behan* (sister) or dharma-*bhai* (brother) is still practiced today in north India.

In the Hebrew Bible, the widowed Ruth loves her mother-in-law so much that she vows never to leave her, insisting that only death can separate them from each other. This famous vow is incorporated into many Jewish and Christian wedding ceremonies, both cross-sex and same-sex. In 1985, a bishop of the Syrian Orthodox church united two American women scholars in a rite of spiritual sisterhood at the Holy Sepulchre in Jerusalem. The bishop said this tie was stronger than biological sisterhood, and would last beyond the grave.[41]

The modern Indian imagination still bears traces of premodern traditions that understand marriage as a type of friendship and friendship as a type of marriage. Thus most Hindu wedding ceremonies incorporate an ancient metaphor for normative friendship—that of seven steps taken together or seven words spoken together.

Seven Steps/Words Together: Marriage as Friendship

The *saptapadi* or seven steps taken together and seven verses recited together, is an important Hindu wedding rite. Each step is taken while a Vedic verse is recited that exalts a particular dimension of marriage, such as wealth or energy. Children are not mentioned in these seven verses, which focus on marriage as an end in itself, not a means to procreation. The seventh step, which completes the ritual, is taken for friendship. In the accompanying verse, the bridegroom addresses the bride as *sakha* (friend) and asks her to become his *saptapada*, that is, one who has taken seven steps and recited seven verses along with him.

This refers to the idea, reiterated in many ancient texts, from epics to fables, that seven steps taken together or seven verses recited together constitute friendship (*saptopadam hi mitram*).[42] This phrase depends on the double meaning of the word *pada*, which refers both to a "step" and to a "verse." Once two people have taken seven steps together or recited seven verses together, they find that they have become friends.

Like the disciples who walk with the risen Christ to Emmaus, conversing all the way, but recognizing him only at the end of the walk, one may be surprised into friendship. This happens in the *Panchatantra*, when a crow proposes friendship to a reluctant mole, and after an extended conversation about the possibility of an interspecies friendship, the crow tells the mole that since they have spoken seven verses together, they now are friends, an argument the mole accepts.

Through this ritual, Hindu marriage is constructed as a type or subset of friendship, which, in the Rig Veda Samhita, is the overarching relationship binding all beings together. In the final verse, the groom also tells the bride to become his *anuvrata* or devoted follower. As in Aristotle's schema and in the seventeenth-century Puritan ideal of companionate marriage inherited by the modern West, the wife is a friend but a subordinate friend. When my spouse and I spoke this verse together, we each asked the other to be her *anuvrata*.

Garland and Ring: Love and the Signs of Marriage

Both in India and the West, same-sex couples who do not have a public ceremony often exchange rings in token of union. Ring-exchange, central to most weddings in the West, is also popular in India. Rings are generally exchanged, along with other jewelry, at Indian engagement ceremonies. A ring as token of love and marriage is central to the story of Shakuntala.

But the primary signifier of Hindu marriage is an exchange of garlands. Like rings, garlands are circular and symbolize perfection; also like rings, they signify enclosing, binding, and connecting. Indians use garlands on ceremonial occasions to honor guests. In most Hindu weddings, the bride's mother and other relatives also garland the bridegroom. The garland, being circular, echoes the *mangalsutra*, the *tali* chain, and the red, gold or conch bangles, tokens of marriage worn by many Hindu wives and some Christian ones.

In the medieval Eastern Church too, bride and groom wore garlands and crowns. Garlands represent amorous union in some Western texts; for example, in Keats' 1819 romantic poem, "La Belle Dame Sans Merci," the lover weaves flower garlands for his beloved's head, arms, and waist. In Indian Muslim and Christian weddings, elders garland bride and groom.

In north India today, exchange of garlands is culturally perceived as central to *gandharva* marriages that are based on love. This is because in most north Indian Hindu weddings, the bride's act of garlanding the groom signifies her choice and consent to marry him. In drama, especially Hindu religious drama, exchange of garlands, along with walking round the fire, often serves as a shortened version of a wedding, and when garlands are exchanged, the audience applauds.

It is therefore not accidental that several women who married each other in recent years signaled their weddings by the exchange of garlands, which features prominently in newspaper reports and photographs. Policewomen Leela and Urmila (see photo 5.1) went to a photo studio in Mhana and had photos taken of themselves garlanding one another. The garlands were preserved in Urmila's family home, hanging on pictures of the Gods.[43]

According to other reports, they also exchanged vows and garlands in a temple at Sagar, in a *gandharva* ceremony conducted by a Brahman. Ramlal Tiwari, a family friend who was at the wedding, remarked: "the marriage was a very low-key event with only 40 people present during the ten-minute religious ceremony."[44]

Reporter Chinu Panchal, notes, ". . . they got themselves photographed garlanding each other. Such a photograph is equivalent to marriage in their parlance."[45] As discussed in chapter 2, photographs often constitute primary legal evidence of male-female marriage too. When a jealous battalion mate stole Leela and Urmila's photographs and handed them to the commandant, the women were first imprisoned and subjected to physical examination and then suspended from the police force and forcibly transported to Urmila's village, on the pretext that they were a bad example to other policewomen.

The two women were harassed so much that, although in February 1988, they and their families had described their love and marriage to several

Photo 5.1 Leela Namdeo and Urmila Srivastava. Photo: *Savvy*.

journalists, by May of the same year, they had accepted the version put forward by police officers and denied that they had ever gotten married or had any romantic or physical relationship. Interviewed separately by *The Illustrated Weekly of India*, in May 1988, both said they were just friends and were having fun when they got themselves photographed as a couple. They wanted to apply for reinstatement, and police officers said the request would be favorably considered. Clearly, economic and social pressure forced the women to lie about their relationship.

Yet, the police authorities had taken the marriage seriously enough to discharge the women from their jobs. In the West too, opponents of same-sex marriage view religious same-sex weddings with the same uneasy mixture of recognition and nonrecognition. This is because it is hard for any society to entirely disregard the socially intelligible deployment of recognizable symbols of marriage.

Aspiring to Marry

Before same-sex marriage became controversial in Euro-America, literary texts represented same-sex couples using tropes of marriage to express their emotions. Poet Gertrude Stein (1874–1946) refers to her companion of 40 years, Alice Toklas, as her "wife" in several poems, and Radclyffe Hall (1880–1943), in her classic lesbian novel, *The Well of Loneliness* (1928), shows her heroine Stephen, like Hall herself, yearning to marry her woman lover.

Modern Indian literature too shows couples exchanging vows, rings or garlands, drinking from one cup, and dying together; it also shows female couples dressing as bride and groom, and taking photographs together, which, as discussed in chapter 2, real-life couples do in token of marriage. In Bankim Chandra's Bengali novella, *Indira* (1893), two women playact the roles of husband and wife, culminating in a passionate kiss; in Nirmala Deshpande's Marathi story, "Mary had a little lamb," (1982), Shobhana and Rose make vows of lifelong friendship, feel that they were attached in a former life, and get themselves photographed dressed as bride and groom; in Ambai's Tamil story "One Person and Another" (1997), Arulan commits suicide after Matthew, his lover of 40 years, dies.[46]

In R. Raj Rao's novel, *The Boyfriend* (2003), Yudi and Milind dress as bride and groom for a private wedding ceremony. Bihzad, hero of Kunal Basu's *The Miniaturist* (2003), paints himself and Emperor Akbar, whom he loves, engaged in "the rites of a secret betrothal."

The Romantic Context of Same-Sex Love

The women who married each other or committed suicide together in recent years met and developed their relationships in different venues—at religious functions, at school or college, or at work. In this, they were no different from cross-sex couples in India today. Since dating is not widely approved of in India, a marriage based on love or attraction typically results from a chance encounter through family friendships or work relationships. Semi-clandestine dating may follow. For example, Neeru, who married Meenu in 1993, first heard her sing at a worship session (*jagran*), and went up to praise her singing. After that, Neeru would drop by on her scooter to take Meenu out, and also give her a ride to work.

Same-sex love relations often develop in the same romantic environment as do cross-sex relations. For example, Lalithambika and Mallika, who tried to commit suicide together in 1980, met when they were Pre-Degree students at Sree Kerala Varma College (SKVC) in Trissur, Kerala. Here is a description by Reji G., male alum of this coeducational college, of romance at this college: "Many students first taste love when they join the college. SKVC had its own set of Romeos and Juliets. These people roamed around the college campus, regularly visited the college canteen and would be in a world of *their* own. . . . The maximum number of lovers are usually found in the Pre-Degree class rooms."[47]

Leela and Urmila's love also grew in an atmosphere saturated with romance. They were in an all-women battalion with a hundred members. Leela and Urmila told a reporter that often two women in the battalion would become very close. "Then we would say *joda bana liya* (they have become a couple). A good many of them have paired off in our company." Two women constables had fought for the favor of a third. Policemen had made advances to Leela and policewomen to Urmila. In this context, Leela and Urmila could not understand why such a fuss was made about their marriage in particular.[48]

Same-sex relationships, like cross-sex relationships, often develop from childhood friendships. In the paradigmatic Indian movie romance, *Devdas*, the lovers, Devdas and Paro, are childhood friends and allies. Many of the female couples who married in recent years were friends from childhood. Same-sex friendship, unlike cross-sex friendship, continues with family approval through adolescence and young adulthood. For example, Raju and Mala continued their cross-caste childhood friendship into adulthood. Their families violently disapproved of their marriage and Raju's mother stated, "We had no idea that the two were up to other things than being just good friends. They were always seen together. We used to laugh when they used to say they would get married to each other."[49]

Are Festivals of Love Indecent?

The Indian debate today about the legitimacy of romantic love is encapsulated in the debate about Valentine's Day, which has become increasingly popular in urban India over the last decade, among both cross-sex and same-sex lovers.

The Shiv Sena (supported by other right-wing organizations like the RSS and the ABVP) has declared a ban on Valentine's Day celebrations, claiming that public displays of romantic love are un-Indian and indecent. Jumping on the bandwagon of antiglobalization nationalists, Shiv Sena leader Uddhav Thackeray declared that Valentine's Day "is nothing but a Western onslaught on India's culture to attract youth for commercial purposes."[50] These same organizations also condemn same-sex relationships as Western, and contrary to Indian culture and tradition.

In the last few years, Shiv Sena activists in Bombay, Delhi, Lucknow, and other cities have violently attacked commercial establishments on Valentine's Day. In 2001, even though no government ban was declared, Hareshwar Patil, Bombay mayor and also a Shiv Sena member, announced, "If any such indecent celebrations are held, they will not be allowed."[51] In 2003, Shiv Sena activists organized public bonfires of greeting cards, broke windows, and harassed handholding couples.[52] The Shiv Sena used similar violent tactics to attack theaters showing films with lesbian content, such as *Fire* and *Girlfriend*.

This is ironic because long before Valentine's Day was heard of in India, festivals of love were held in honor of Kama, God of love, and his male companion, Vasanta (spring). The descendants of these festivals are still celebrated today—Holi, in north India, with its Saturnalian and erotic content; Thiruvathira, in Kerala, a women's festival dedicated to Kama; Pooram, also in Kerala, when Kama is worshiped; and Saraswati Puja, all over India, when Kamadeva is worshiped with the Goddess of wisdom, education, and the arts.

In medieval and modern India, Vasant Panchami or Basant, the spring festival, direct descendant of Vasantotsava, is an occasion for all types of boundaries to be crossed. *Haqiqat al-Fuqara* (1662), the Persian poetic biography of Sufi mystic Shah Hussayn, recounts how Hussayn, a Muslim, fell in love at first sight with Madho Lal, a young Hindu man but was unable to meet him. Finally, during Basant celebrations, when Hindus and Muslims mingle, Madho, singing and dancing, poured colors on Hussayn, and they grew intimate. Overcoming Madho's Brahman family's hostility and censorious Muslim authorities' persecution, the two lived together and were inseparable. They never married women, and Hussayn became known as Madho Lal Hussayn.[53]

Ironically, the Shiv Sena, which is virulently anti-Muslim, takes exactly the same view of Valentine's Day as do Muslim extremists. Lawrence Wright, who spent a year in Saudi Arabia, reports that the religious police, accompanied by the state police, spent the week before Valentine's Day "going through card and flower shops, attacking anything that was red or had hearts on it; florists hid their roses as if they were contraband."[54]

Hindu right-wing organizations are out of tune with mainstream Hindu traditions in this regard. Despite the Shiv Sena's efforts, Valentine's Day continues to be celebrated. "It's our day," Kamna Saxena was quoted as saying, while she and her boyfriend Nitin Sharma had coffee in a crowded restaurant. "How can anyone be against love?"[55] Stores sell Valentine's Day merchandise, prices of roses soar, magazines and newspapers run Valentine's Day cover stories on February 14, and hotels offer Valentine's Day specials.

Is Romantic Love Good?: Cultural Ambivalence

In the United States, opponents of same-sex marriage claim that it will undermine the institutions of marriage and the family, and destroy civilization itself. They sublimely ignore the fact that heterosexual marriage and family are rapidly changing, quite independently of same-sex relations, for example, fifty percent of marriages in the United States now end in divorce, and the majority of Americans find remarriage after divorce completely acceptable.

In India, where the legitimacy of romantic love as the basis of marriage is itself in question, the debate is whether romantic love undermines the social fabric or sustains it. Those Indians who see love as an antisocial force point to the high divorce rate in the West to back up their claim that marriages based on romantic love are less likely to last than are family-arranged marriages. If longevity is viewed as the only measure of a good marriage, they are probably right. Many other Indians, however, do not agree that longevity is the only measure of a good marriage. They claim that individual choice and temperamental compatibility (as distinct from familial compatibility) are equally or even more important.

On the surface, most Indians might appear to fear romantic love and most Westerners to approve of it. In fact, though, Western and Indian traditions of romantic love overlap considerably. Also, the view of love as antisocial is more widely prevalent in the West than is apparent at first glance, as I argue in chapter 1.

This ambivalence is evident even in the Saint Valentine stories. According to one legend, he was a priest who performed marriages, defying the Roman Emperor, who had forbidden young men to marry because he thought unmarried men made better soldiers. According to another legend, while Valentine was in prison, he wrote a letter to his beloved signed "Your Valentine," thus inaugurating the tradition of sending love letters on this day. Some historians believe that Valentine's Day is a Christianization of the ancient Roman fertility festival of Lupercalia. Valentine is recognized as a saint because he acted for the welfare of others and to promote love rather than war. Yet it is significant that in all the legends he was martyred. Websites of evangelical Christian churches in the United States emphasize that Valentine's Day should be an occasion for married couples to renew their Christian commitment to each other. This anxiety to contain love within monogamous heterosexual marriage is very different from the idea of love as an uncontrollable force that is found in ancient and medieval Western cultures.

Ideas of romantic love are deeply ambivalent both in Indian and Western cultures. The argument for same-sex marriage is primarily based on an appeal to the social legitimacy of romantic love and coupling, and on a less obvious appeal to the idea of marriage as friendship. Proponents of same-sex marriage claim that stable couples promote social stability and thus contribute to the larger good. Heterosexual lovers in modern India often appeal to similar ideas of romantic love in order to gain acceptance. Families often accept heterosexual love marriages when the couple produces a child. Today, many same-sex couples in the West are producing and raising children; some raise children in India too. In the next chapter, I further explore how same-sex parenting intersects with ideas of same-sex marriage.

Chapter 6

Monstrous to Miraculous—Same-Sex Reproduction and Parenting

If Americans are so repulsed by gay sex, perhaps the solution is to just allow gays to marry and have kids. After all, everyone knows that parents of young children have no time for sex.

—Columnist Gersh Kuntzman, Newsweek, *August 11, 2003.*

Most cultures perceive children as a divine blessing bestowed on individuals, couples, families, and society. Although this blessing is ideally supposed to connect with love between a child's parents, children are frequently the product of indifferent, loveless, even hate-filled connections between a man and woman. Whether or not it is the product of love, the child is (or was, until recently) a product of sexual intercourse between a man and woman, for which the word "love" is often a euphemism. It is on the basis of this euphemistic connection between sexual activity and "love" that antigay forces claim that heterosexual "love" is blessed with fertility while homosexual relationships are cursed.

I suggest that cultural ambivalence regarding non-goal-oriented individual pleasure, of which sexual pleasure is symbolic, is resolved in the cult of children—children are perceived as the justification of heterosexual pleasure. This justification often spills over to homosexual pleasure too. Contrary to the right-wing extremist view, voiced by the Vatican among others, that gay parenthood constitutes violence to children, many same-sex couples who have children report that heterosexuals tend to view them much more kindly as parents than as lovers: "Parenthood trumps gayness."[1]

Are children a blessing bestowed only upon heterosexual couples, or can they be a blessing on same-sex love as well? The Hebrew Bible tells of a child who has a father and mother, but is symbolically the son of two women, Ruth and Naomi. Ruth, Naomi's widowed daughter-in-law, is deeply devoted to her. She obediently marries the man her mother-in-law picks out for her. When Ruth has a son, he is seen as the son of Naomi too: "And the women said unto Naomi, Blessed be the Lord which hath not left thee without a kinsman, . . . for thy daughter-in-law, which loveth thee, which is better to thee than seven sons, hath born him. . . . And Naomi took the child, and laid it in her bosom . . . And the women her neighbours gave it a name, saying, There is a son born to Naomi" (Ruth 4:14–17). This is no ordinary child—he is the ancestor of King David. This story, like that of Bhagiratha's birth to two mothers, suggests that where love abounds, blessing can be unexpectedly abundant too.

Children of Same-Sex Parents

Historically, when same-sex couples raised children, these were usually the biological children of one partner, often from a former cross-sex marriage. In the West today, many same-sex couples decide to have children together. They may adopt, but generally face serious discrimination in the adoption process. Some countries, like Netherlands and Sweden, allow same-sex couples to adopt; others, like Belgium and France, do not. Others, like the United Kingdom, allow a single gay person to adopt, but not a couple. This means that if the adoptive parent dies, his/her partner will have no parental rights, and the child's future is jeopardized. In the United States, the law is unclear and varies from state to state. Same-sex couples are often among the few willing to adopt disabled babies or older children with disturbed histories. In the United States, they also foster children whom the state finds hard to place; sometimes, after they have cared for these children for years, a homophobic state official decides to take the children away.

In India, some adoption agencies have recently formalized guidelines denying gay couples right to adopt. Unwanted children grow up in orphanages but same-sex couples are not allowed to adopt them. However, private arrangements are often made. Same-sex couples or gay individuals adopt children from low-income families. In many cases, the biological family remains involved with the child, even though the adoptive parents are the primary caregivers. One female couple adopted a neighbor's child after its mother died and its father remarried. Another female couple adopted a child who worked as a domestic servant.

Hijras too raise adopted children. Tikkoo, a *hijra* who is a Muslim and has worked in the film industry for decades, found a 12-day-old baby abandoned on the roadside, her cheek bitten by a rat, and worms crawling out of her stomach. He adopted her and named her Tamanna. She is now grown up but still has the scar from the rat bite on her cheek (see chapter 9 for a film based on this story).[2]

Another *hijra*, Mona Ahmed, also a Muslim, relates how she got her daughter, Ayesha, as a four-day-old baby.[3] Ayesha's father abandoned her mother during the pregnancy, and when the grandmother asked Mona's advice, Mona offered to adopt the baby. She paid for the mother's expenses during pregnancy, but never met her. The mother died in childbirth, and the grandmother gave Mona the baby. Mona lived in a house with four other *hijras* and their guru, Chaman, also a *hijra*. Although Ayesha's family had given her to Mona, the whole *hijra* family raised Ayesha. When Ayesha was four, Mona became an alcoholic, and quarreled with the family. While Mona was out of town, the family relocated and took Ayesha away. Since the adoption was informal, Mona had no legal claim to the child. She continued to meet Ayesha, but lost custody of her.

In the West, many same-sex couples now also choose to produce children. Sometimes, unfortunately for an overpopulated world, they are pushed into reproducing because discriminatory laws prevent them from adopting. Female couples use artificial insemination. Sometimes, one partner is impregnated with sperm from a relative of her partner's, so that the child will be biologically related to both. Male couples use surrogate mothers—the sperm of one or both partners is used to impregnate a woman who is paid for her services. These are techniques widely used by infertile male-female couples in India as well. But same-sex couples in India have not begun to use these techniques, because of the heavy stigma still attached to a woman having a child outside marriage.

Selfish and Unselfish Pleasures

Parents often respond to a child's coming out as gay with the accusation that s/he is being selfish. One Indian mother asked her U.S.-based gay son, "Don't you want to give anything back to society?" by which she meant give children to society.[4] The accusation has larger ramifications. It connects to the unstated but persistent idea that heterosexual couples somehow "pay" for their pleasure by making sacrifices for their children.

In the Judeo-Christian-Islamic tradition, this idea may stem in part from God's curse of Eve. God tells Eve that she will suffer in childbirth. Many

European Christians took this curse so seriously that when anesthesia was first invented, they objected to its use in childbirth, because they believed women's labor pains were divinely ordained and should not be mitigated. That curse and sacrifice are inseparable from blessing is clear in the Hebrew Bible, when Rachel, who obsessively longs for children and resorts to many stratagems to have them, dies in childbirth and names her baby "Ben-oni," meaning "son of my sorrow" (Genesis 35:18).

What pleasure or profit are parents supposed to derive from children? In most societies, including India, children, especially sons, constitute parents' old age insurance. A son enhances a widow's status—he enables her to stake a claim in her husband's family property, and he is also supposed to support her in old age. In the West, with the development of welfare benefits, medical insurance, retirement plans, and senior citizens' homes, adult children are less likely to live with and support their parents. This difference is well encapsulated in the fact that government employees in India can declare their retired parents their dependents and put them on their health benefit plans, whereas this is not the case in the United States.

A more complicated pleasure is that of seeing one's own self immortalized in one's child.[5] In Plato's *Symposium*, Socrates' teacher Diotima argues that males and females of all species engage in sexual love because they are seeking immortality through their offspring. Antigay thinkers today would stop here, and claim that same-sex couples, deprived of this immortality, engage in infertile love. Diotima, however, argues that those persons who are more fertile in mind than body (a tendency she restricts to the human species) produce children of the mind (books, ideas, institutions), and same-sex lovers fall into this category.

We can complicate Diotima's argument by noting that human parents, unlike parents of other species, are not usually content with children's physical resemblance to them. They try to mould the children's minds and personalities, thus aspiring to turn children of the body into children of the mind.

Most cultures recognize the role of other adults—spiritual parents, teachers, godparents—in the shaping of children. Hindu traditions sometimes place the teacher or guru above parents in this respect. Both Indian and Western literary traditions are replete with examples of inadequate biological parents being replaced by loving adoptive parents.

When same-sex parents raise children today, the non-biological parent, although often legally disadvantaged, may shape the child as much as or more than the biological parent. A female couple I know in India raised the two children of one partner from her former marriage to a man. One child physically resembles the biological mother but in temperament and taste is recognizably the child of the other mother.

Sacrificing of and for Children: Abraham and Others

Do heterosexual people really sacrifice more for others than homosexuals, assuming that more heterosexuals produce children? If one's child is a reproduction of one's self, then sacrifice for one's child is not really sacrifice. Even giving up one's life for one's child, commonly cited as the ultimate example of maternal love, involves not just ending one's life but extending it through the child's.

That is why medieval Christian stories show a person who sacrifices his children to save his friend as the exemplar of Christian virtue. Children are not really "others," hence one's love for them is more egoistic than altruistic. A friend is truly an "other"; thus sacrificing one's children for a friend is truly altruistic. When Amile, at the angel's suggestion, kills his children to heal his friend's leprosy, he imitates God's sacrifice of his son for the human race, and also Christ's sacrifice of himself for his "friends" who are not biological kin: "Greater love hath no man than this that a man lay down his life for his friends."

A similar idea is found in ancient Hindu stories of men who sacrifice wife, children, and themselves to satisfy the needs of a guest or a supplicant. Such is the story of Prince Srilaya, whose father is a Shiva devotee. To test him, Shiva, disguised as a hermit, demands that the king kill and cook his son to feed him. The king does so, and the boy is miraculously restored to life.

This premodern idea of virtue has not survived. Today, most Christians would see Amile's sacrifice of his child for his friend as morally repugnant. Abraham's near-sacrifice is much more attractive to modern Christians, Jews and Muslims. The focus in this story is not on the child's suffering but on Abraham's suffering. Abraham suffers because he has to deprive himself of that which is most precious to him. Although Abraham sacrifices not for his son's sake but for God's sake, the dominant feeling in the story is that of suffering parenthood. Victorian parents who punished children to the accompaniment of the famous phrase, "This hurts me more than it does you," were modeling themselves on Abraham.

Abraham is generally figured as a righteous man suffering for no fault of his own. However, pursuing my earlier suggestion, I would argue that Abraham suffers at least in part to pay for his pleasures. His relationships with his wife Sarah and concubine Hagar have already resulted in suffering for both women and for Ishmael, his son by Hagar. Within the psychological economy of the story, which recurs in many narratives of sex and reproduction, Abraham's own suffering is necessary to justify his continuing pleasures (after Sarah's death, he marries again, though he is then over a hundred years old).

While parents are perceived as unselfish because they give up pleasures while rearing children, this unselfishness is questionable because childrearing is often experienced as pleasurable—giving birth to and nursing the baby Jesus are counted among Mary's seven joys, not her seven sorrows. Devotional songs celebrating the babies Krishna and Rama depict their parents lost in delight at their first toddling steps and childish pranks.

Childrearing for Society?

Perhaps the only definitely unselfish reason for having children is to contribute to a depleted population. Few people in the world today can claim to have children for this reason, except perhaps when particular communities, like the Parsis, the Jews or some tribal peoples, find their numbers shrinking. Historically, though, producing children within certain lineages, rightly or wrongly perceived as socially important, has often been seen as a duty rather than a pleasure. Thus, Henry the Eighth's obsession with producing a son was shared by many of the British, who feared that the ending of the Tudor line would lead to warfare and chaos.

In the Hebrew Bible as in many ancient texts, the anxiety to produce children, especially sons, is related to the need to reproduce a threatened human population in general, and one lineage in particular. God's blessing "Be fruitful and multiply," first at creation, and then after the flood, is not bestowed in the context of overpopulation, but in that of an empty world.

Nevertheless, most societies today, whether over or underpopulated, still tend to see fertility as a divine blessing. A child is *felt* (rather than thought) to be produced not just by its parents but also by some third force, whether nature with a capital N or God or the Gods. Any greeting card store in any city in the world provides evidence that the rhetoric of miracle is more popular than the rhetoric of nature—cards congratulate parents on the blessing or miracle of a child, not on the biological outcome of their heterosexual intercourse. Why, then, when two women produce a child together by artificial insemination or two men through a surrogate mother, do some in the West see this reproduction as monstrous rather than providential or miraculous?

Female-Female Sex and Reproduction in Ancient Medical Texts

The ancient Hindu medical text, the *Sushruta Samhita* (ca. AD 100–100 BC) attributed to the sage Sushruta, states that a boneless child, (interpreted by commentators as a child with cartilaginous bones) is the result of an act of sexual intercourse between two women, in which their *sukra* or sexual fluids unite in the womb of one of them.[6] This statement is part of a list of gender and sexual deviations diagnosed as medical conditions.

According to the *Sushruta Samhita* as well as a contemporaneous medical text, the *Charaka Samhita*, conception occurs due to an aggregate of five elements—father, mother, the Self (*Atman*), suitability, nutrition, and mind.[7] Of these, the Self is the most important, as it causes not only birth in a particular species, but also mind, sense organs, respiration, consciousness, memory, ego, and will. The *Charaka Samhita* states that when the maternal element preponderates the child is female, when the paternal preponderates the child is male, and when both are equal the child is of no sex or what today would be called intersexed. The mother contributes the baby's skin, blood, flesh, fat, and all the fleshy organs such as heart, liver, kidneys, stomach, and intestines. The father contributes the baby's hair, nails, teeth, bones, veins, and semen.

According to Sushruta, a woman dreaming of sexual intercourse can conceive and give birth to a jelly-like mass. Sushruta suggests cures for some other conditions, but does not prescribe any cure for babies born without bones. In the context of pathologizing sexual differences, this text seems to suggest that such births are monstrous, and the result of impure acts.

This ancient idea of the possibility of two women reproducing was not forgotten. Some sacred texts produced in Bengal from the fourteenth century onwards wrote it into a devotional context, turning the monstrous into the miraculous. Several elements are involved in this process of transformation, the most important being the rewriting of sex acts in the context of love, and the envisioning of female-female sexual love in a particular context as divinely blessed.

Bhagiratha's Birth to Two Women: A Story Unique to Bengal?

The story of Bhagiratha's birth to two women occurs, as far as I know, only in texts produced from the fourteenth century onwards in Bengal, eastern India. Among these texts are Bengal manuscripts of the *Swarga Khanda* section of the *Padma Purana* (in Sanskrit but written in the Bengali script), and various versions of the *Krittivasa Ramayana* in Bengali.[8] Most other texts relate that Bhagiratha was born in the regular way to his father, Dilipa, and, as is standard in patrilineages, do not even mention his mother.[9]

The *Krittivasa Ramayana* is "said by literary historians to be the most popular single book in all of premodern Bengal; and . . . has retained its popularity" today, with half a dozen editions available in Calcutta book markets.[10] Bhagiratha is an important figure to Hindus because he brought the sacred river Ganga to earth from heaven, a task his father and several forefathers had failed to perform. Bhagiratha is also the ancestor of epic hero Rama, incarnation of the preserver God Vishnu.

One of the names of the river Ganga is Bhagirathi, in Bhagiratha's honor. His name survives today in the popular imagination through the expression idiomatic in several Indian languages, *Bhagiratha prayatna* (referring to the feat of bringing down the Ganga) which is the equivalent of the English "Herculean feat."

Why does the story of Bhagiratha's birth to two women appear only in texts produced in Bengal? I suggest that this is in part because of medieval Goddess worship traditions. Bengal was and is one of the centers of Goddess worship and also of Vaishnavism. These two traditions often assume syncretic forms in Bengal. The Bengal texts that tell the story of Bhagiratha's birth to two women are primarily Vaishnava texts, glorifying Vishnu, but I suggest that Goddess worship traditions influence the way these texts develop the idea of two women procreating.

The God of Love and the Sage of the Kamasutra: The Story in the *Padma Purana*

The *Padma Purana* is an accretive text, one of the eighteen major *Puranas* (compendia of stories and rituals celebrating different Gods and Goddesses) compiled between the fourth and the fifteenth centuries.[11]

The narrator of the story in the Bengal recension is the primal serpent, Sheshanaga, and the interlocutor is sage Vatsyayana. Neither of these characters appears in this story in the standard version of the *Padma Purana*. Vatsyayana is famous as the putative author of the *Kamasutra*. Via this allusion to the *Kamasutra*, in conjunction with an implied reference to the medical text, the *Sushruta Samhita*, the Bengal *Padma Purana* places itself in the context of sacred textual traditions, both medical and erotic, that are already at this time over a millennium old. It claims the authority of those traditions to legitimize its account of Bhagiratha's miraculous birth, and also develops those traditions by interpretation, giving a new slant to the idea of female-female sex and procreation.

In the Bengal *Padma Purana*, after king Dilipa dies childless, his two widows visit the family priest Vasishtha in his hermitage and request him to help them continue the family line. They need a child to bring the Ganga from heaven so that rituals can be performed to redeem Dilipa's ancestors. Vasishtha meditates, reaches a resolve, and says, "I foresee that you will have a great and auspicious son, so I will try my best to fulfill your purpose."[12] He then performs the *putreshti* sacrifice, which is still performed today for married couples who want a son, and prepares specially cooked rice. Giving it to the queens, he tells them that one of them should eat it and the other should have sexual intercourse with her, with the *bhava* of a man (*purushabhavena maithunaya*).[13] No physical change takes place in the queen who takes on the *bhava* of a man, so the term suggests that the queen perform a desire that is usually attributed to a man.

The queens follow his advice and the older one becomes pregnant. The text does not give any description of the women's lovemaking or emotions. The child, conceived without semen, is born without bones (interpreted by commentators as having cartilaginous bones), and is named Bhagiratha because he is born of "the *bhaga* (vulva) alone": *Bhagiratheti tannama jato yad bhagmaatratah.*[14] This explanation for the name Bhagiratha, found only in the Bengali script texts, is important because it explicitly credits the female reproductive system with independent generative power.

Here, the use of the motif of the magical food and the lack of any description of the women's relationship subordinates their lovemaking to the issue of male lineage. Their relationship is motivated and sanctified only by the need to continue the male line, under the pressures of duty to male ancestors. Even so, it is significant that, although many other forms of miraculous birth are available in the ancient epics, this text represents Bhagiratha's birth as resulting from the sexual union of two women.

Although Bhagiratha, being boneless, is crippled and ugly, he grows up and learns all the Vedas in his childhood. On the way to study with his teacher Vasishtha, he one day encounters the sage Ashtavakra, whose name literally means "bent in eight places." Ashtavakra, as his name indicates, is deformed, and when Bhagiratha salutes him, the sage suspects that the boy is mocking him by mimicking his crippled condition. Enraged, the sage declares that if the boy is mocking him, he will be burnt to ashes, but if he is naturally crooked, he will immediately attain beauty and strength. Thereupon, Bhagiratha becomes a strong youth, as beautiful as Kama, God of love. After thanking Ashtavakra, he proceeds to meet Vasishtha, who is so impressed by his beauty that he crowns him king.

The Story in the *Krittivasa Ramayana* A

The Bengali Ramayana attributed to the poet Krittivasa who lived in the fourteenth century, is also an accretive text—additions continued to be made to it up to the eighteenth century.[15] There are several versions of the *Krittivasa Ramayana*, and the story of Bhagiratha appears only in some manuscripts. In the version most widely available in Bengali today, which has also been translated several times into English, the story is an expansion of the one in the Bengal *Padma Purana*.[16] As a translation of the excerpt has appeared elsewhere, I summarize it here instead of reproducing the translation.[17] For purposes of comparative analysis, I refer to this version as *Krittivasa A*.

After king Dilipa dies childless, it is not the widows but the Gods who grow worried because Vishnu is to be incarnated as Rama in Dilipa's lineage. Now that Dilipa is dead, the line seems to have come to an end. The Gods hold a consultation and send Shiva to Ayodhya. Shiva goes to Dilipa's two widows and tells them they will have a son by his blessing. When the widows ask how this is possible, Shiva instructs them to have sexual intercourse with each other.

The text provides a short description: "The two wives of Dilipa took a bath. The two young women lived together in extreme love [*Sampritite achhilen se dui yuvati*]. After some time, one of them menstruated. Both of them knew one another's intentions and enjoyed love play [*keli karitey*], and one of them conceived."[18]

The child is born as a boneless lump of flesh. Ashamed and grief stricken, the queens decide to throw him into the river Sarayu. Vasishtha intervenes and advises them to leave him on the roadside instead. After they leave, Ashtavakra sees and curses the deformed child who gets transformed exactly as in the Bengal *Padma Purana*. The two queens are delighted and take their son home. Ashtavakra performs the child's naming ceremony, naming him Bhagiratha because he was "born of two vulvas" (*bhage bhage janam hetu bhagirath nam*).

Growing Up as the Child of Two Mothers

The next section in *Krittivasa A* is, as far as I know, unique among medieval texts for its depiction of the problems faced at school by a child of two mothers. Parts of the description resonate with the predicament of children of single mothers and lesbians today.

When Bhagiratha is five years old he is sent to study at Vasishtha's hermitage. One day, when the children who study there are quarreling with each other, another child calls him *jaraj*, literally a child born of a mother's lover (*jara*, ancestor of the modern word *yar*). Bhagiratha is deeply hurt and makes no answer, but does not return home. With tears in his eyes, he lies down in the sulking room (*kopagraha*) at school. His mother grows worried and, like a tigress deprived of her cub, asks Vasishtha where her son is. She embraces the child, wipes his tears, and promises to find a doctor to cure whatever ails him.

Bhagiratha then asks his mother, "To what lineage do I belong and to what clan, and whose son am I?"[19] His mother tells him the entire story and also tells him he was named Bhagiratha because he was born of two vulvas. Hearing this, Bhagiratha laughs.

When his mother asks why he is laughing, he replies that bringing Goddess Ganga to earth is not a small task but a "Bhagiratha" feat that only he can perform, and which he will now set out to do. His two mothers try to dissuade him because he is so young, but he does not listen to them. He takes *diksha* (initiation) from Vasishtha and bids his mothers goodbye. He then performs severe austerities, addressed to each of the Gods in turn, which ultimately result in his bringing the Ganga to earth.

Chandra and Mala's Monsoon Romance: *Krittivasa Ramayana B*

In another version of the *Krittivasa Ramayana*, found in only one manuscript, which I shall refer to as *Krittivasa B*, the two queens' personalities are more fleshed out.[20] They have names, and rather than being instructed to make love, they do so spontaneously, inspired by the God of love who inspires all lovers, the son being an unexpected and initially unwelcome by-product.

Here is the relevant section, for the first time translated into English:

Ashamanja's son was Anshuman,
and his son was Dilipa. Listen all in the name of Rama.
There was no greater king on earth.
Chandra and Mala were his youthful brides.
One day the king went to hunt deer

but could not see even one.
Disappointed, he began his retreat but
then he spotted a pair of deer copulating.
Unmindful of the consequences the king shot an arrow
and killed the male deer while in the act of mating.
Grieving her mate's loss, the female deer cursed the king:
"May you die, if you go near a woman."
Thus cursed, he returned to his kingdom,
but was unable to enter his palace on account of the curse.
The king stayed away from his palace
and managed the affairs of the State from there.
The king lived thus for awhile
but one day, weak with desire, he stepped into his palace.
He embraced Malavati and
true to the doe's curse he died.
The king's body was lifeless
and the women grieved inside the palace.
The kingless kingdom became unruly
and civil life was disrupted.
The gods in heaven congregated to address the matter.
Without the Suryavansha [the sun's lineage] the world would cease to be.
Brahma, Vishnu and Maheshwara [Shiva]
met on Mount Kailash.
Lord Brahma conferred with all the Gods
and decided to summon Sage Vasishtha.
He requested Vasishtha to help Chandra and Mala get a son.
"Vishnu Vishnu," said the sage, covering his ears,
and refused to comply with their wishes.
After Vasishtha's refusal
they called upon Madan [God of love].
Brahma directed Madan to make haste
and make a son be born from the stomachs
of Chandra and Mala.
Obeying Brahma's bidding Madan
went straight into the inner quarters of the palace.
As Madan reached the king's palace
the two queens began menstruating.
Three days later they took the purifying bath,
entered their husband's palace, and lay down there.
The sky was overcast with clouds,
the swans sang and the peacocks danced.
The skies darkened and a stormy rain followed.
Burning with desire induced by Madan, Chandra and Mala
took each other in embrace,
and each kissed the other.

Chandravati played the man and Mala the woman
[*Chandraboti purush hoilo Mala hoilo nari*]
The two women dallied and made love
[*Dui nari mono ronge rongo krira kori*].
God's blessing had enabled the two women to play the game of love
and the energy [*tej*] of Madan [god of love/desire] entered the womb of Malavati.
This is how Malavati became pregnant.
The news was welcomed with sounds of joy.
One, two, three, four, five, seven of her friends
teased her about her pregnancy.
When Malavati realized she was pregnant
she began to weep and say:
"I have not been with a man -
how can I be pregnant?
I will be denounced as a woman of ill repute.
So I will drown myself in the Sarayu river."
She stepped into the Sarayu to end her life
but Brahma stopped her.
"Listen carefully to me," he said,
"Unless the lineage of Suryavansha carries on,
the world itself will be lost.
Lord Narayana [Vishnu, incarnated as Rama] must be born in your line of descent,
which is why the Gods had to plan this strategy.
They sent Madan to your inner chamber,
to make it possible for two queens to make love.
It was Madan's energy [*tej*] that impregnated you,
so the son in your womb will be beautiful.
If there is any demerit (*paap*) within you,
let me bear it and you can go home free of it.
Your son will be the incarnation of God
and his able hands will save the world."
Content, the queen returned to her palace
and ten months passed.
In an auspicious hour a beautiful boy was born.
The fair skinned boy rapidly grew,
his beauty was unequalled in the three worlds.
Six months later, his naming ceremony was conducted,
following the instructions of holy books.
Born of mutual enjoyment between two vulvas [*bhage bhage sambhog*],
He was named Bhagiratha by God Brahma
He read the fourteen Vedas and at the age of twenty,
became king and ruled his people.[21]

Love Made This Possible

In chapter 1, I suggested that the emotional context in these narratives enables the transformation of the potentially monstrous (sex between two women leading to childbirth) into the miraculous. What, then, are the emotions that animate these texts?

The framing emotion is devotional love, directed toward the Gods, in the *Padma Purana* especially toward Vishnu, and in the *Krittivasa Ramayana*, especially toward Rama. Within this frame there are other emotions—the queens' unselfish concern for their husband's lineage, kingdom, and ancestors; their love for one another and their child; the Gods' concern for the universe and for humans; and the sages' concern for the royal family, the people, and the Gods. The text draws the reader in to share all these emotions. On the other hand, there is the conflicting emotion of distaste at possible pollution caused by sex between the two queens. The texts resolve these conflicting emotions by privileging some emotions over others.

All of these are Vaishnava texts, that is, they are animated by love for Vishnu, the preserver God. Bhagiratha's birth is part of a larger divine plan for the preservation of the universe. Rama, the incarnation of preserver God Vishnu, is to be born in the royal line of Ayodhya. The Gods bless the two women's relationship because it furthers that plan. For the devout reader or listener the possibility that Rama might not be incarnated or the Ganga might not come to earth is a fearful one. The reader is drawn into sharing the Gods' anxiety and is reassured that the plan for the women to become lovers must be good because the Gods devise it.

I suggested earlier that the only reason for having children that is arguably entirely unselfish is to replenish an underpopulated community. The birth of Bhagiratha occurs in precisely that type of context. The ending of Dilipa's lineage endangers not only Dilipa himself, since he has no son to perform his last rites, but also all of his ancestors, whose last rites cannot be performed until one of their descendants brings the Ganga to earth.

Most important, Ganga's coming to earth is seen as essential to the material and spiritual welfare of the human race. In the *Padma Purana*, Dilipa's widows are motivated by a selfless anxiety about these matters; they nowhere express a selfish desire that would be understandable—for a son who would redeem them from their lowly status as childless widows.

Resolving Negative Emotions

Although Hindu traditions do not invest non-normative or *ayoni* sexual acts with anything like the horror with which Christian traditions invested "sodomy," Hindu ideas of pollution inform all sexual relations with some degree of distaste. In addition, Bhagiratha's mothers are widows. The visceral anxieties associated with pollution, non-procreative sex, women's sexual pleasure, and widows' sexual pleasure, are confronted in these texts and resolved through other emotions.

In both the *Padma Purana* and *Krittivasa A,* this anxiety takes the form of the child's deformity at birth. This deformity is cured when the child turns out to be virtuous. Though he is the son of two mothers, or rather because he is the virtuous son of two virtuous mothers, Ashtavakra heals him.

In *Krittivasa B,* negative emotions surface most powerfully, and are most clearly resolved. The first negative emotion is Vasishtha's horrified response when the Gods suggest that he help the two windows have a child. In the *Padma Purana*, Vasishtha matter-of-factly instructs the two women to make love. But in *Krittivasa B,* he embodies mainstream Hindu asceticism's opposition to sexual pleasure, especially widows' sexual pleasure. However, the Gods ignore Vasishtha's horror, and carry on regardless. The chief divine agents here are creator God, Brahma, and God of love, Kama. The principle of creative love overrules that of asceticism.

The next negative emotion is Mala's panic when she finds herself pregnant. While in *Krittivasa A,* the child Bhagiratha is teased at school, here his mother is teased and fears disgrace. In *Krittivasa A,* the baby Bhagiratha is almost drowned in the river; here, his mother almost drowns herself.

This introduces the element of suffering central to the image of a child as a blessing that validates sexual pleasure. Mala pays a price for her pleasure but the Gods tell her the price is not too high. In the other two texts, the two women pay the price when the child is born deformed, since they suffer extreme grief until he is cured.

When Brahma appears to Mala, to stop her committing suicide, he gives a name to all these anxieties. He says, "If there is any demerit (*paap*) in you, let me bear it and you can go home free of it." The word *paap*, common in modern Indian languages, is often loosely translated as "sin," but this translation is a Christianization. *Paap*, is in fact, closer in meaning to "impure/demeritorious action," as opposed to *punya* or pure/meritorious action. *Paap* in the Hindu context is very different from sin in the Christian context. *Paap* is demerit born of bad actions that attaches to the self and causes rebirth; however, the merit born of good actions also attaches to the self and causes rebirth. Demerit will result in a lower birth and suffering while merit will result in a higher birth (even a birth as a God or demi-god) and happiness. In Christianity, since the soul is born only once and after death is either saved by faith or damned by sin, sin is potentially much more deadly.

Brahma's offer suggests that there may be some demerit or impurity associated with the queens' love making. He does not elaborate on this suggestion nor does he say that there definitely is impurity associated with it; (he uses the conditional "if"). The impurity in question could be related to same-sex relations; more likely, it could refer to the taboo on widows indulging their desires. Brahma takes the possible impurity on himself; he also indicates that the impurity, inspired by the God Kama and counteracted by him, is not a major one. The *Kamasutra* declares that objects normally considered impure are pure for certain purposes, for example, although saliva is impure, a woman's mouth is pure during sex.

Divine Blessing Overrides Medical Proscription

The texts are pervaded by the emotion of wonder directed toward the Gods. In the *Krittivasa Ramayana* (as distinct from the *Padma Purana*), the Gods' direct intervention and blessing sanctifies the two women's relationship. In a sacramental understanding of marriage, whether Christian or Hindu, it is divine blessing that sanctifies a secular relationship. It is also divine blessing that makes the apparently impossible possible.

In *Krittivasa A*, when Shiva tells the women they will have a child, they ask, "We are widows, how can we have a child?" He replies, "By my blessings one of you will have a lovely child." One may compare Mary's question, "How shall this be, seeing I know not a man?" and the angel's reply, "With God, nothing shall be impossible."

Shiva is a God associated with gender transformation, varying eroticisms, and miraculous birth. He is connected to femaleness through his *ardhanarishwara* (half-man, half-woman) form, and to homoeroticism through his playful transformation into a female to please his wife Parvati in loveplay.

In *Krittivasa B*, Brahma tells the two women that the God of love made their lovemaking possible. The Gods' agency and blessing is more powerfully evident in *Krittivasa B* than in the other texts—here, the inauspiciousness of a deformed birth is preempted by love. What is this love? It is an amalgam of the different types of love celebrated in the text—devotional love, romantic love, maternal love, familial love, all embodied in the God Kama.

The presence of Kama trumps both the medical prognostication that a child born of two women's union will be boneless and also the impurity possibly associated with two widows' sexual union. The extended romantic description of the women's love, attraction, and sexual union in this text is, then, not fortuitous but directly relevant to the auspicious outcome.

The etymology given for Bhagiratha's name is the same as in the other texts, but there is an added stress on sexual pleasure, and it is a God, not a sage, who names the child:

> Bhage bhage sambhog je tathe upagata
> Brahmadev thuilen nam bhagiratha.
> (Born of mutual enjoyment between two vulvas
> He was named Bhagiratha by God Brahma.)

The word "*sambhog*," used in this version but not in the others, literally meaning "mutual enjoyment," is the word generally used to signify sexual intercourse even today. The *Padma Purana* mentions only one *bhaga*, *Krittivasa A* mentions two, and *Krittivasa B* mentions the pleasure shared by two *bhagas*.

Widows' Sexual Pleasure: The Triumph of Kama

That the two women are widows is a significant fact. Krittivasa's *Ramayana*, a normative sacred text in Bengal, endorses widows' sexual pleasure, thus flying in the face of the stereotype that Hindu widows, especially in Bengal, are stripped of agency and forbidden to indulge in pleasure, especially sexual pleasure. The widows' pleasure surfaces in the interstices of this patrilineal narrative, overflowing into an excess of sensuous description—the conventionally romantic monsoon season, the kisses, the burning desire, the presence of Kama.

In the *Padma Purana*, the God of love is mentioned only as a simile—the healed Bhagiratha is said to be as beautiful as the God of love. But in *Krittivasa B*, he is an important agent. My argument is that the Gods, especially Kama, overcome anxieties regarding the legitimacy of sex between women.

Krittivasa B overcomes these anxieties by deploying several strategies. If pleasure is socially taboo for widows, the taboo may arise from the awareness that, as sexually experienced women, they are more likely to engage in sexual activity. Conversely, the taboo on a virgin's having sex is stronger because it destroys her marriageable status. The widowed status of the two women may paradoxically enable their intercourse, because the anxieties demonstrated in the *Manusmriti* regarding virginity and marriageability (see chapter 1) are irrelevant to widows. Also, having sex with another woman may be perceived as less illegitimate than having sex with a man who is not one's husband (although, under certain circumstances, that too happens with divine blessing in the ancient epics).

Having experienced marriage to a man, Dilipa's widows continue to fulfill the aims of the patrilineal family, yet they also find love and pleasure in union with one another. This is perhaps the most important validating factor—as co-wives, they are required to love one another, and may even in a sense be married to one another. In chapter 7, I further explore co-wifehood and other kinship structures as sites for same-sex love.

As the son of two widows (rather than a widow and another man) Bhagiratha's paternity is represented as unambiguous. He is his father's son because he is born to his father's wives. By convention and law, (as was also the case in most Euro-American laws, prior to DNA paternity tests), the child of a Hindu married woman is considered her husband's son unless anyone can prove otherwise. In conversations with the Gods, including Shiva, Bhagiratha introduces himself as the son of Dilipa, not of the two queens. The fact that the queens are Dilipa's wives ensures his legitimacy. A Hindu widow who produces a child would conventionally be subject to censure and persecution, and her child regarded as illegitimate. The text constitutes Bhagiratha as an exception to the rule—like Jesus, he is represented as having no human male for a father but is nevertheless called the son of his mother's husband.

Is Female-Female Sex Impure?

Unlike Mary, Bhagiratha's mothers engage in an act of sexual intercourse in order to reproduce. How is it that these sacred texts accommodate and even celebrate a same-sex sexual act which appears to be anti-normative and violative of prescriptive texts like the *Manusmriti* and the *Arthashastra*?

My answer is: because it is *ayoni* or non-vaginal sex that is prohibited in Hindu law books. And Bhagiratha, far from being a product of *ayoni* sex, is, so to speak, a product of double-*yoni* sex ("*bhage-bhage*"). The clue lies in the texts' odd insistence on the folk etymology of Bhagiratha's name—born of two vulvas (*bhagas*).

In *Krittivasa 1* there seems to be some awareness that this etymology is suspect, for the text asserts the putative author's reputation as a scholar immediately after providing the etymology: "Because he was born of two vulvas (*bhaga*) he was named Bhagiratha. The great poet Krittivasa is a recognized scholar (*pandit*). In this Adi Kanda he sings the birth of Bhagiratha."[22]

The type of intervaginal intercourse described in the Bhagiratha texts is nowhere mentioned in the Hindu law books. As discussed in chapter 1, it is manual-vaginal penetrative intercourse between two women when one of them is a virgin that is proscribed in the *Manusmriti*, and intercourse between non-virgin women is not mentioned. The *Manusmriti*'s concern is for the loss of virginity and the consequent unmarriageable status of the girl, thus a virgin who manually penetrates another virgin is supposed to be punished with a fine and whipping, and also the payment of double the penetrated girl's bride price, while a mature woman who does it to a virgin is supposed to have her head shaved and two of her fingers cut off (8. 369). This is the most severe punishment prescribed for any form of same-sex intercourse in the Hindu law books. But exactly the same punishment, having two fingers cut off, is also prescribed for a man who manually penetrates a virgin (8. 367). This punishment, then, is not for same-sex intercourse, but for the act of taking a girl's virginity, thus imperiling her chances of marriage.

By repeating the word for vagina (*bhaga*) the Bhagiratha texts enact and underscore the female-female intercourse that results in miraculous birth. The idea of the primal and pure fecundity of the Goddess appears to hover behind this construction. It is reinscribed in Bhagiratha's feat of bringing the Ganga to earth. The Ganga, herself a Goddess, is a purificatory and salvific force.

Bride and Groom?

Let us list the ways Bhagiratha's mothers' relationship is represented as a marriage-like union:

1. As co-wives, they are in a lifelong union with one another.
2. In the *Padma Purana*, the family priest performs a *putreshti* sacrifice (to get a son) for them, before they have sexual relations. This sacrifice is normally performed, even today, only for a husband and wife.
3. In *Krittivasa A* and *B*, the Gods plan, bless, and oversee their sexual relationship. The match between them is literally made in heaven.
4. Like the women whose marriages in India have been reported since the 1980s, one woman acts as a wife and the other as a husband.
5. Their relationship results in the birth of a legitimate child, who is heir to the kingdom.

Miraculously Born of Desire

Gods and heroes in most mythologies are conceived and born miraculously—from virgins, from human-divine intercourse, or from a single parent, male or female. The miracle functions to signal the hero's innate difference from other mortals. As John Boswell has shown in another context, heroes are also often raised differently from other children—by adoptive or foster parents, human, divine, or animal.[23] This may signify that they belong not to one family alone but to the whole society.

In Hindu texts, one of the most common forms of miraculous birth is a God, demon, or Goddess producing other beings from the self. When this happens in battle, these beings are born of wrath and are terrifying. They aid the parent in fighting. In the *Devi Mahatmya*, the Goddess, when invoked by the Gods to destroy the demons, creates an army of divine female beings from herself. A similar phenomenon occurs in most Goddess texts. This may be seen as a type of cloning because these females are embodiments of the Goddess's different attributes; they mirror her and may merge back into her.

Sometimes, however, a female produces another being not from wrath but from other emotions such as erotic or motherly love. Thus, in the *Padma Purana*, Vishnu, disguised as the demon Jalandhara, seduces Jalandhara's wife Vrinda. While they are engaged in love play, Tulasi, a purifying nymph, arises from Vrinda's sweat. Tulasi (identified with the sacred plant, holy basil) represents Vrinda's pure erotic desire for Vishnu.[24] Goddess Parvati produces the God Ganesh from her body rubbings because she wants a son of her own, who will be devoted only to her.[25]

Goddesses' ability to produce fully formed beings appears to be related to the idea, also found in ancient Greek texts, that the earth (also a Goddess) produces certain types of life, such as worms, from herself. Sita, heroine of Valmiki's epic *Ramayana*, springs from the earth and finally returns to the earth.

Parthenogenetic Goddess as Human Woman

Bhagiratha's mothers, however, are not Goddesses. They are ordinary human women. How, then, do they participate in a type of reproduction generally reserved for Goddesses or other divine beings?

Some feminist critics argue that Hindu worship of Goddesses has no positive effect on women's status, because Hindus view Goddesses as totally different from women, and Goddesses do not share women's suffering.[26] In my view, although Goddess worship does not have a one-to-one equation with improving women's status, it is not true that Hindus see Goddesses and women as totally different. In fact, the Goddess is seen as latent in every woman, and her powers become manifest under suitable conditions.

On certain occasions, such as the festival of *Kumari Puja* in north India, and also at birth and during the wedding ceremony, human females are especially charged with divine energy. This is reflected in the practice of referring to daughters, wives, and daughters-in-law as Lakshmi, Goddess of prosperity. Goddesses do not always function to empower women but sometimes they do.

Because Goddesses, like Gods, are sometimes born as humans, the line between Goddesses and women is further blurred. The best example of this blurring is the epic heroine Sita, who represents both women's suffering and women's resistance. From the *Valmiki Ramayana* through the many other versions of the Rama story produced over two millennia to versions being produced today, Sita has consistently been a catalyst for Indian debates on the injustice of men toward women. Hindus worship Sita as simultaneously human and divine, and Hindu women identify with her sufferings as well as her powers.

In the *Adbhut Ramayana*, a fourteenth-century Sanskrit version of the Rama story, Sita is born parthenogenetically from a woman, and later assumes a terrifying form in order to slay a demon much more powerful than the demon killed by her husband, Rama.[27] Rama, who was not aware of her true identity as a Goddess, humbly worships her as the supreme principle, creator, protector, and destroyer of the universe. After granting him a vision of her universal form, Sita resumes her human form. These events are also depicted in some Bengali Ramayanas of the same period.

Goddesses' destructive powers are inseparably linked to their reproductive powers. Both are specifically female powers and therefore not entirely divorced from human women's abilities.

Beyond the Order of Nature

Like the birth of many heroes, Bhagiratha's birth is not in "the order of nature." But most cultures acknowledge at least two ways of being unnatural—a phenomenon may be supernatural or divine, or it may be subnatural and demonic. Bhagiratha's birth, like Jesus', is framed as supernatural.

When males in Hindu texts reproduce miraculously, a woman or at least an apparent woman is always involved as the inspirer of desire who causes the ejaculation—the sages see beautiful women and ejaculate; Shiva sees Vishnu in the form of Mohini and ejaculates to produce Harihara. But when a Goddess produces autonomously, she can do so without involving a male.

I see these ideas as imaginative envisionings of the detachment of reproduction from penile-vaginal intercourse. Like Leonardo da Vinci imagining the airplane long before it could be constructed, the Bhagiratha narratives imagine women producing a child together. I would invoke reality here—many co-wives rear children together, and some have sexual relationships with each other. These texts imagine what would happen if such female partners could produce a child together.

Mark Jordan argues that anti-gay thinkers' insistence that homosexual sex is proscribed because it is non-procreative arises from the fact that the Christian West has not fully eschewed its historical condemnation of erotic and sexual pleasure in general: "The irrational force of the Christian condemnation of Sodomy is the remainder of Christian theology's failure to think through the problem of the erotic."[28] Jordan points out that many branches of modern Christianity, in their celebration of families and reproduction, have "degenerated into fertility cults," thereby giving up the Gospels' prioritization of spirit over body. Writing as a Christian, Jordan sees the celebration of biological fertility as pagan, not Christian.

One might also point out, however, that a "pagan" emphasis on biological fertility in conjunction with an acceptance of desire and bodily pleasure as fundamental to life might be congenial to the construction of same-sex desire as potentially, if miraculously, fertile. As discussed earlier, Hindu ascetic traditions developed a deep suspicion of bodily desire and pleasure, but this suspicion always was and still is contested in mainstream Hinduism by the dominant idea of Kama or desire as one of the four normative aims of life. Such a concept of bodily, this-worldly pleasure as a major life-goal is not to be found in mainstream Christian theology.

The blessing of same-sex intercourse with a miraculous child in the Bhagiratha texts may be read as a heterosexist assimilation of same-sex coupling; it may, conversely, be seen to function as an affirmative incorporation of same-sex sexual and amorous relationships within a religious norm of the sanctified life.

Co-Mothering as Cultural Ideal

Families with a father and mother are not the only ones idealized in ancient texts. Families raised by co-mothers, like Kunti, Madri, and their sons in the *Mahabharata*, are also idealized. Heroes often are privileged to have more than one mother figure—many medieval and Renaissance paintings envision Jesus as reared by his mother Mary and grandmother Saint Anne.

In the Bhagiratha texts, the two women who miraculously produce a child together raise him together; conversely, two males who miraculously produce a child rarely raise it together. This is perhaps because male-male births are unplanned and unexpected. When Shiva and Vishnu (in the form of Mohini) together produce a son named Harihara (Hari = Vishnu; Hara = Shiva), Mohini is embarrassed and abandons the child on earth where he is found and adopted by a childless royal couple, and grows up to become the God Ayyappa. In a later legend, Ayyappa, when questioned as to his relationship with Parvati, wife of Shiva, and Lakshmi, wife of Vishnu, becomes puzzled, and retreats to the forest where he is still worshipped.[29]

However, single men are shown as tender adoptive fathers., For example, Shakuntala, born to a sage and a heavenly nymph, is abandoned by her biological parents and raised by another sage who finds her and adopts her as his daughter. Thus, non-biological parenthood of various types emerges in Hindu sacred stories as both a social reality and an ideal.

Co-mothering by pairs or groups of mothers appears as an ideal as early as the hymns of the Rig Veda (ca. 1500 BC). Agni, God of Fire, one of the most important deities in the *Rig Veda*, is repeatedly described as "child of two mothers"(*dvimatri*), and occasionally, "child of three mothers" (the three worlds).[30] His two mothers are sometimes the two sticks from which fire is generated for the sacrifice. The gender of the firesticks (*arani*), in Sanskrit, is feminine. The lower *arani* is laid flat, and the upper *arani* is rapidly rubbed against it. Friction, not penetration, generates fire.[31]

These ideas of multiple mothering and female energy, arising from ancient sources, including the Vedas, the epics and the Puranas, form a palimpsest that help us place in context the Bengal versions of the Bhagiratha story.

Families of Love

The miracle of Bhagiratha's birth to two women is not an easy one—it involves conflict, struggle, and defiance of social norms. Only the Gods consistently support Bhagiratha and his mothers. One may compare the medieval English plays that dramatize both Joseph's rejection of Mary as adulterous and her vindication by the angel of God.

One may also compare the struggle of same-sex couples today to raise children together, as seen in the story of Sheela and Sree Nandu. Sheela had an affair with a male cousin, who abandoned her when she became pregnant. Her father beat her severely, and threw her out of the house. She took refuge in various orphanages. When her friend Sree Nandu, who was also unhappy in her family, learnt of this, she went and lived with Sheela in an orphanage, and nursed her through the delivery and various illnesses. As they did not have anywhere to live, they left the child in the orphanage.

Now that they have been enabled to live together (see chapter 4 for details), Sree says, "We want to adopt a child. We don't know how to reach the child that Sheela gave up for adoption, but would like to give any child a home and build a family quite different from our families. You don't need to have your own children. And what is the guarantee that all parents will love their children? We know that from our own lives. Sheela and I want to give shape to a family where love alone would be supreme."[32]

Several gay and lesbian Indians in the West have created such families of different types, ranging from children raised in marriages of convenience, to children produced by a lesbian couple with insemination by a gay male friend, to adopted children.[33]

Like Bhagiratha, these children's parents make possible the seemingly impossible. Their existence blesses not just their parents and lineage but also society and humanity at large. Bhagiratha's feat of bringing the Ganga down to earth is a fit symbol of this blessing.

Chapter 7

All in the Family: Same-Sex Relationships in Traditional Families

Anyone for a marriage of convenience? A young and handsome gay man . . . industrialist from India . . . very sensitive, caring, honest and in the closet, . . . seeks a companion for a platonic relationship. Would prefer to have a baby.

—*On khush chat-list, 2004*

Love's secret is always lifting its head
Out from under the covers,
"Here I am!"

—*Jalaluddin Rumi (1207–1273)*

Perhaps the best-kept secret about homosexuality is that as many if not more homosexually inclined people worldwide live in "traditional" heterosexual marriages as in same-sex couples. Those who are traditionally married never get counted as gay in surveys. It is well known that in India and other supposedly traditional societies, large numbers of people live as apparently traditional heterosexuals, while secretly engaging in homosexual liaisons, or leading lives of quiet desperation. That the same is true in the West is less often acknowledged because many people assume that the openly gay community is synonymous with the entire gay population. In fact, this is far from being the case.

Opponents of same-sex marriage sing the praises of the traditional family not acknowledging that the traditional family, is not and never has been as homogenous as it appears. A newspaper report filed on December 10, 2004, which happens to be Human Rights Day, told the story of Venu, a 40-year-old auto rickshaw driver in Kochi, Kerala, who, on the insistence of his 26-year-old wife, Mangala, to whom he has been married for ten years, agreed to marry her coworker and long-time female lover, Ramlath. Mangala wanted to live with both of them and threatened to commit suicide if she could not have Ramlath live in the house. Venu's second marriage took place at the temple in Guruvayoor. Ramlath's family complained to the police, and Venu with his wives and two sons by Mangala, moved to an undisclosed location.[1] In this chapter I further explore co-wifehood and other kinship relations as sites for female-female love.

Puritanical forces would like to imagine that same-sex relations exist only in seamy underworlds outside of respectable society; they cannot endure the revelation that same-sex love is everywhere, including in the heart of the family. In 1927, Hindi novelist Pandey Bechan Sharma "Ugra" published a collection of eight stories about homosexuality entitled *Chocolate*. The book sold out immediately, but was virulently attacked by the literary establishment, including such supposed liberals as the writer Premchand, because Ugra, while claiming to denounce homosexuality, depicted homosexuals not as slum or prison dwellers, but as respectably married middle-class men, both Hindu and Muslim, with flourishing social networks of their own, engaging in liaisons in their homes and in public spaces, such as schools, colleges, and poets' gatherings.[2]

Of course, even larger numbers of traditionally married people secretly engage in heterosexual liaisons. This is termed adultery, but homosexual extramarital relationships are not always treated as equally serious infringements. In 2003, the Supreme Court of New Hampshire held that a woman who is married to a man and has sex with another woman is not committing adultery because adultery requires penile-vaginal intercourse that could result in offspring.[3]

Governmental refusal to validate same-sex marriage does not put an end to same-sex relationships or strengthen the traditional family. On the contrary, by refusing equal rights to gay people, governmental discrimination functions as an incentive for gay and bisexual people to enter into heterosexual marriages and lead double lives, in accordance with well-established "tradition."

Double Lives Worldwide

In the 1980s, public health agencies working on HIV and AIDS in the United States coined the term MSM (men who have sex with men) for apparently heterosexual men who do not term themselves bisexual or gay but who regularly have sex with men. In the last few years, some organizations working on HIV in India have turned this health-related term into a sexual identity term, claiming that Indian MSM are an indigenous category, more authentic and less Westernized than Indian men who identify as gay (see Note on Methodology). Many people on South Asian chatlists, usually espousing a feminist perspective, challenge these HIV activists' nonjudgmental view of men who have homosexual sex outside marriage and often transmit HIV to their wives and children. These activists in turn accuse feminists of using a Western framework to judge an indigenous reality.

In fact, there is nothing indigenous or uniquely Asian or Indian about homosexual men who are married to women and lead a double life. They are very common in the West too.[4] In the United States, African-American men who have sex with men but see themselves as heterosexual have coined the phrase "on the down low" or "DL," to describe their sexual arrangements. This phenomenon is now attracting attention due to the higher rates of HIV transmission among African Americans. But this high rate is caused largely by poverty, not by higher rates of male-male activity. Male-male sexual activity is likely to be equally high among the white and other American populations.

African American novelist E. Lynn Harris brought the "down low" phenomenon to mainstream attention in the 1990s in his best-selling novels. His trilogy, *Invisible Life* (1994), *Just As I Am* (1995), and *Abide With Me* (1999), traced the growth of protagonist Raymond Tyler from living as heterosexual while being on the DL to coming out as gay and living openly with a male partner. In a semiautobiographical work of nonfiction, *On the Down-Low*, J.L. King discusses the phenomenon and its huge impact on HIV transmission.[5]

Novelist Edmund White depicts a similar phenomenon among white men, in his novel *The Married Man* (2000), where an American man falls in love with a Frenchman married to a woman. Earlier depictions of this phenomenon among both black and white men in the United States include James Baldwin's fiction, such as *Another Country* (1992).

As Shohini Ghosh points out, terming such men MSM makes sense only if we term men who have extramarital sex with women MSW![6] So-called MSM are homosexual and bisexual men. Some of them find casual partners in cruising spots, such as bars, restaurants, streets, and parks. A gay male Indian friend told me that in any crowded public space in India numerous men make eye contact with one another, which most women would not notice or understand. They often follow up this eye contact by going off together for a quick encounter.

On the other hand, many men married to women also have long-term relationships with men. A remarkable relationship of this kind is recorded in the documentary film *Terhi Lakeer* (Crooked Line, 2002). Two elderly middle-class Hindu businessmen in Delhi have been in a relationship for thirty years. One of them, Harish Agarwal, tells the interviewer that he met his friend a couple of years after his marriage. They became close and developed a physical relationship. Agarwal says that he is "110 percent gay," and that had he met his friend before he married he might not have married, because all one needs in life is love. When Agarwal's lover separated from his wife, Agarwal and his family persuaded the couple to reconcile, so that he would have children to look after him in old age. The two couples lived together for about ten years, raised their children together, and are still very close. The men say they are like brothers and their wives like sisters. So close are the families that when Agarwal's son was to be married, the other couple made all the arrangements. Agarwal says their families are not aware of the sexual dimension of the men's love, and that he does sometimes feel anxious about this being discovered. Agarwal says he is very proud that their love has lasted so long.

The parallel female phenomenon of women who are married to men but have relationships with women on the side tends to be overlooked, because women generally meet each other not in public but in private spaces and thus are less visible. They also attract much less attention from health agencies as they are less at risk for HIV. In the early 1980s, a male social worker from Rajasthan visiting *Manushi* told me that rural men who have to travel on business often prefer their wives to have relationships with other women than with men, because the former do not result in pregnancy. In feminist circles in India in the late 1970s and 1980s, many women, some married to men and others single, maintained liaisons with other women while passing for heterosexual. Some of the leading activists in virtually every autonomous women's group were engaged in such relationships. Some later came out as lesbian, but many did not.

How "Traditional" is Traditional?

Proponents of the traditional family in the West today assume that a man and woman romantically in love with and married to each other constitute a "traditional" family. In fact, this type of unit is less than a century old even for the majority in Euro-America and is therefore not very traditional at all. Before that, most marriages were at least in part family-arranged, and other considerations besides romantic love determined matches. Family-arranged marriages are even today the norm for most of the world's population, including significant groups in the West, such as many Orthodox Jews,

Mormons, Asian and African immigrants, and some Christians too, especially in rural southern Europe.

Biblical "traditional" marriages are arranged by parents. Such is the normative marriage of Isaac and Rebekah. A marriage based solely on romantic love, like that of David and Bathsheba, or Esau and his non-Jewish wife, violated the "traditional" norm and was disapproved of. Jacob's marriages to his first cousins, although based in part on romantic love, were approved because marriage between first cousins was an approved type of marriage.

This chapter focuses on two of the many ways same-sex relationships exist in the traditional family. One is through kinship. As an example of kinship, I examine the institution of co-wifehood. Despite the horror some right-wing Christians in the West express toward polygamy and co-wifehood, these are Biblically sanctioned institutions, which, whether we like it or not, constitute a highly "traditional" type of marriage in many societies, including some minority communities in the West and in India.

The second institution is even more interesting because it is simultaneously "traditional" and modern. It is a type of marriage many Indian gay men and lesbians are now arranging for themselves—a non-sexual arranged marriage between a gay man and woman, which appears entirely "traditional" and heterosexual to everyone except those in the know. This type of marriage is the product of modern sexual identity construction (the partners see themselves as "gay," "lesbian," or "bisexual"), and is aided by international networks and the internet. However, it grows directly out of modern monogamous marriage based on individual choice.

These kinds of marriages too are not unique to India or Asia. They were known as "front marriages" in the 1950s in the United States, and they occur today as well. In her memoir, lesbian historian Lillian Faderman describes her "front marriage" with a gay male friend in 1958.[7]

Wanted: A Nonsexual But "Traditional" Spouse

In the West it is no longer acceptable to openly espouse the idea of marriage as a purely social arrangement entered into as a duty, to please one's parents and society. The idea of marriage as a practical arrangement was effectively killed off in Euro-America in the early twentieth century. While many Europeans and Americans do still marry for practical reasons, almost no one is willing to acknowledge it openly. Everyone feels obliged to make a pretense of romantic love.

In India, the older concept of marriage as a social duty still flourishes. This makes it possible for many gay Indians, even Indians living in the West, to openly search for gay partners of the opposite sex. Such arrangements used to be made through lesbian-gay networks of friends, but now are also made in South Asian gay media and websites. While some gay Indians condemn such arrangements as hypocritical, others consider them dutiful rather than cynical.

I studied the Personals in twenty issues of *Trikone*, the magazine for lesbian, gay, bisexual and transgender South Asians, published from San Francisco. The issues ran from 1998 to 2003. I found that 11.5 percent of all Personals were advertisements placed by gay people looking for gay people of the opposite sex to enter into a marriage of convenience (MOC). Another two percent were placed by "traditionally" married bisexual or gay people, looking for same-sex relationships on the side.

There are at least two websites and a chat-list entirely dedicated to MOCs for South Asian gay people. MOCs are also discussed on other South Asian gay chatlists, leading to heated debates.

The term "marriage of convenience" used to refer to heterosexuals who decide to get married for financial, social, and practical rather than romantic reasons. But today it is generally used disparagingly, and therefore almost never occurs in heterosexual advertisements.

The people placing the ads are both male and female (although many more are male). They include Hindus, Muslims, Sikhs, Parsis, and Christians. Most are Indians or Americans of Indian origin, but there are several Pakistanis, and some West Asian Muslims. Most are highly educated professionals. Some live in India and others in the United States and the United Kingdom. Some were born in the United States to Indian parents. Some are willing to relocate if necessary.

Parental and Societal "Pressure"

Most advertisers say they need an MOC because of parental pressure to marry: "Decent, very good looking, 29 years, professional doctor, Pakistani citizen in the U.S., need to get married to a lesbian female due to intense family pressure."[8] Some go into greater detail: "I'm 28, a very good looking Indian (Sindhi) guy. My parents are wondering why I don't have a girlfriend and I am being pressured to get married. I would like to meet a girl (gay) or one who has no problem marrying a gay guy. I am very str8 acting, I can be a great boyfriend/husband. I just have other needs that I would like to fulfil. I'm 28, 5'10, 155, in good shape. Very active, love music/dancing, love outdoor activities. I work in finance."[9]

The advertisers mix lingo from Matrimonials in India, like "decent" and "from a respectable family," with phrases common in Personals in the West, like "hoping to meet my true soulmate"[10] or "funloving, easygoing."[11] Caste, regional, and religious identities are often mentioned but not insisted upon.

The advertisers rarely explain what kind of "pressure" their parents exert. Since almost all are employed and economically independent, living separately from their parents, often even on different continents, the "pressure" is clearly emotional rather than physical or economic. A 21-year-old woman in a long-term relationship with another woman explains the pressure. She says she is "very comfortable" with her sexuality and is "out" to all except her family. Her parents discovered her lesbianism in the past and threatened to sever their relationship with her. Although she does not live with them, she wants to "ensure their happiness" and refrain from hurting them. So she is looking to marry a gay male, comfortable with his sexuality, who would respect her girlfriend and have no romantic expectations from the marriage: "I would only want this to be a working marriage. This means a marriage that appears to be 'working' whenever my parents or family inquire into our lives."[12]

Sometimes, gay couples advertise for MOCs for both partners. Thus an Indian, living in the United States, advertises for MOCs for himself and "a friend," adding that the "ladies' sexuality is not important."[13] An American woman advertises, saying that her partner is an Indian Brahman woman and needs an MOC.[14]

In some cases, the "pressure" seems to be more amorphous than direct. Advertisers refer to social expectations as well as to their own internalized feeling that heterosexual marriage, even if nonsexual, is a desirable arrangement: "I am a 28 closet Indian male looking for a bi/lesbian to spend the rest of my life with. I was born and raised in the United States. For society, family, and personal reasons I would like to build a marriage of convenience."[15]

On a chat-list, a male software professional points out that if one remains single past a certain age everyone understands that one is gay. He argues that this could hurt one's parents and have negative repercussions, perhaps even result in losing one's job.[16] Therefore, he claims, the practical solution is for a gay man and a lesbian to have an MOC, and live together as friends, maintain separate finances, but work out an arrangement whereby each can bring sexual partners home at different times, and be "out" only to close friends.

Some advertisers insist that, unlike the many gay and bisexual people who cheat on their heterosexual spouses, an MOC is honest and mutually beneficial: "I refuse to marry someone without them knowing anything about my situation. I don't want to hurt or destroy someone else's life."[17]

"Traditional" Philosophy of Marriage?

Several advertisers claim that an MOC is consistent with traditional values: "Educated professional gay male looking to find a lesbian partner who would be willing to get into a relationship of convenience so as to get the best of both worlds; and so that the societal pressure of marriage which is such an Asian value can be fulfilled. I strongly believe this has a good potential to enable one who is gay to live a life of convenience and symbiosis so that there is honesty and no heartbreak in a relationship. I am an honest person and would hate to hurt an innocent person by a regular marriage."[18] Another man, looking for a "straight-acting young lesbian of Indian origin" writes, "If you, as a gay Indian female, want to be true to yourself without toppling a delicate world, drop me a line."[19]

A lesbian opponent of MOCs puts forward an argument based on ideas of individual integrity: "Why participate in a farce, which undermines one's self-esteem and foundation of one's character?"[20] A proponent of MOCs retorts, challenging the ideas that character is primarily founded on sexuality and that marriage must be linked to sexuality: "There is a dangerous equivalence of love to sex in the camp that prescribes sexuality-based marriages (or should I say marriage of inconvenience?) for everyone."[21] He argues that all good long-term marriages are based on companionship and shared values, not sexual or romantic feelings. He doubts the likelihood of finding such companionship with another gay man, and therefore considers an MOC a practical way to find lasting companionship.

The vast majority of advertisers say they do not want a sexual relationship with their opposite-sex spouse. A few say they are open to the idea of sex for procreation, and a couple of men say they are bisexual and want a "real" marriage with a woman but also want the freedom to have sex with men on the side. In one unusual case, a 28-year-old self-employed woman living in India says she is "straight" but is "not interested in sex" and hence would like to marry a gay man.[22]

The Will and Grace Syndrome

Several advertisers appear to have more than just practical expectations from an MOC. They seem to be looking for romantic friendship. They aspire toward the ideal of marriage as a type of friendship. A male finance and software professional writes: "I am seeking career oriented, independent person who without obligation of sexual intimacy can become good friend. A person who understands emotions and feelings, has good sense of humor. I expect my partner to be with me throughout my life as best friend."[23] A Punjabi lesbian looking for an MOC with a Punjabi gay man writes, "I'm hoping to meet my 'true soulmate,' purely platonic."[24]

This is an Indian version of the Will and Grace syndrome. One reason "Will and Grace" has proved so popular despite two of its main characters being openly gay men, is that Will's real emotional and romantic relationship is with a woman. That this relationship cannot be sexually consummated paradoxically makes it even more romantic. The reality-TV show, "Boy Meets Boy" also accommodated this syndrome by having the gay protagonist's best friend, a straight woman, be his constant companion on the show, clearly much more emotionally intimate with him than any of the males vying for his favors.

There is something deeply appealing to American viewers, both gay and straight, about a romantic, physically affectionate, and intimate male-female relationship that is normative by virtue of being male-female yet is free of the demands of sexual activity and reproduction, and also free of the social dangers and discomforts of homosexuality. That appears to be part of the lure of the MOC for some Indian people. They seem to think it easier to find nonsexual love in friendship than to find sexual love in friendship.

In a family-arranged Indian marriage, parents look for economic, educational, and social compatibility, but those arranging an MOC for themselves often look for emotional and temperamental compatibility, with shared values. Thus, a British national of South Asian origin who is a marketing manager for a major UK retailer writes that he is gay and very comfortable with his sexuality, and adds: "I am Protestant Christian and very much involved in the Church. . . . The reason for this marriage is to ease the pressure from my parents and to be able to live in peace. I will not expect any physical intimacy, and I would expect the feeling to be mutual. Children will be nice but not a necessity. *The type of person I am looking for is a Professional Female Lesbian, who is under the same pressures as me. I can offer a good life in London, and a loving home. The person must be a committed Christian or someone who will be willing to change.*"[25] (italics original).

Some women want to combine a committed lesbian relationship with an MOC with a gay man; others are open to either possibility. One 33-year-old woman with a Ph.D. even advertises for both simultaneously: "I'm looking for a 'real' relationship w/ a woman or a friendly marriage with a bi-gay/trans man."[26]

Multiple Definitions of "Traditional" Marriage

Like many traditional marriages in pre-twentieth century Euro-America and most traditional marriages in India today, the gay-lesbian MOC is not based on romantic love but is a socially acceptable domestic arrangement. There are many heterosexual marriages worldwide that are not passionately sexual or even sexual at all. Other marriages are sexual only for the purpose of procreating. This type of marriage was, for many centuries, the kind most approved by the Catholic Church, and few Christian churches would entirely disapprove of such a marriage even today, provided both partners remained celibate.

To take a less extreme view, in many traditional marriages, even as late as the nineteenth century in Europe and the United States, husband and wife slept apart in separate beds or even separate bedrooms, especially after they had produced the requisite number of children. This continues to be the case in India. While sharing finances, a home, and social responsibilities, a man and woman may find emotional fulfillment in activities, friendships, or relationships outside the marriage. Marriage can be seen, as Aristotle saw it, as an honorable type of friendship that is not necessarily the supreme type of friendship or love. It can also be seen as a social arrangement through which one fulfils one's obligations to one's parents, extended kinship group, religious congregation, and society. There is nothing opposed to tradition in such marriages.

The major difference between such marriages and the gay-lesbian MOC is that most (but not all) of those entering the latter intend to have sexual relationships on the side. But is even this practice entirely untraditional? Have not many men and some women throughout history conducted discreet liaisons while staying married, and has not "traditional" society deliberately turned a blind eye to such liaisons?

Before the advent of the gay media, gay men and lesbians sometimes found one another either deliberately or inadvertently (while they still considered themselves heterosexual) and were sufficiently strongly drawn to one another as to get married. A famous example is the marriage of British novelists Vita Sackville-West (1892–1962) and Harold Nicolson, both homosexual. They had two sons together, but also had many same-sex affairs on the side. In her private diary, Vita recorded the story of her pre and postmarital relationship with the love of her life, Violet Trefusis. After her death, her son Nigel Nicolson published the diary with a sympathetic introduction, titling it *Portrait of a Marriage*.

I know several male-female couples in the United States who married young, and later both discovered that they were gay. One such couple in Montana has children and a grandchild. They parted amicably and the woman now has a female partner who lives in another town. The ex-spouses remain best friends and have recently become housemates.

Traditional Marriage with Fun/Love on the Side

A marriage of convenience more "traditional" than an MOC is one wherein a man and woman who fall out of love after marriage decide to stay together for the sake of the children or because they are good friends, carrying on affairs, heterosexual or homosexual, outside the marriage, with or without the knowledge of the other spouse. While "tradition" frowns on such arrangements in theory, in practice traditional societies have always accommodated them.

Many Indian gay men argue in favor of this type of arrangement, claiming that they are good husbands and fathers, and what their wives don't know won't hurt them. In most "traditional" societies, whether in Europe or Asia, men had greater freedom to indulge their extramarital sexual desires, as long as they fulfilled their obligations as husbands and fathers. In modern societies, women's assertion of their right to equality, in combination with the new expectation that one's spouse must fulfill one's emotional as well as sexual needs, generally results in a higher divorce rate. That some people today live openly in same-sex relationships is also in part a result both of women's demand for freedom and equality and of people's refusal to sacrifice individual happiness to social obligation.

The internet is now making more visible a phenomenon that used to be well hidden. Many American self-defined "bi-married" men look for other men on America Online, Gay.com and many other websites. These men often describe themselves as only "incidentally" bisexual or homosexual. Some of them say they would never go looking for men in gay bars or cruising spots, which might be unsafe. They claim to be primarily sexually attracted to women and happy with their wives, but are interested in a little safe fun on the side with other men. They consider this a "guy thing" that they do with their "buddies": "On AOL there are chat rooms with names for every particular kind of taste and then some, from "BiMarriedItaliansNYC" and "WifeSleepingNextToMe" to "LosAngelesStr8M4M" and "Str8GuysLookToo," where men swap photos and cell phone numbers."[27]

Like the defenders of such behavior in India, these men claim that their sexual encounters with other men do not constitute "cheating" on their wives. Some tell their wives about their "buddies" but many do not.

Indian married men also advertise on gay sites for sex or relationships with men. Many use the terms "relationship" or "friendship" while others say they want "fun." One, advertising in *Trikone*, states that he is not looking for romance but just for sex. But some married men say they want a gay long-term relationship.[28] One man describes himself as "happily married" and says he is looking for a "straight acting" man for "regular encounters."[29] Almost all specify that they are looking for "discreet" or "straight-acting" men. Some say they would prefer other married men.[30]

The Green Card Marriage

The U.S. government's refusal to recognize same-sex relationships for purposes of immigration means that many gay people, in order to live with their same-sex partners, are forced to resort to cross-sex marriages. The odd paradox is that the U.S. government will recognize for immigration purposes, any marriage, loveless or otherwise, between a man and a woman, but will not recognize a decades-long loving relationship between two men or two women.

In 2004, I edited a special issue of *Trikone*, the South Asian LGBT magazine published from San Francisco, on the theme of immigration. I heard many heart-wrenching stories of same-sex interracial couples separated temporarily or permanently, or forced to migrate to Canada or to live as illegal immigrants in the United States. All of them said they had considered cross-sex marriage at some point during their ordeal.

The Personals in *Trikone* show interesting configurations arising from this unjust situation. Some gay people hold out the prospect of sponsorship as a lure for a prospective spouse. Some opportunists, who may or may not be gay but who want to migrate to the United States, take advantage of the situation by offering themselves as spouses to American passport holders who are gay and need an MOC for familial reasons.

What About Women?

Most women have less power and mobility than most men, therefore married women often find it more difficult than married men to have their cake and eat it too. I found only one married woman advertising in *Trikone* for a same-sex relationship, and she sounds as if she is searching for love rather than fun: "I am in east SF bay, in my mid 30s, an Indian bi woman. I am married, however I miss being with a woman, the touch, the talking, kissing, and the emotional part. P.S. No Men."[31]

Rather than meeting partners in public spaces or cruising spots, wives usually find partners in the family—among relatives, family friends, neighbors, and other associates. In one case of a married woman's liaison with a single woman in Delhi, disaster resulted when the husband found out and threw a fit. In most cases, however, the women manage things with discretion so that the façade remains relatively undisturbed. I also know American married women (both Indian and non-Indian) who have relationships with female neighbors, sisters-in-law or friends, without their husbands' knowledge.

Because the patriarchal family structure generally favors men, women's relationships often exist in the interstices of kinship. In the early 1980s, a woman who came to *Manushi* for help told me the story of how her mother-in-law fell in love with her. The two women spent most of their time together at home, and the mother-in-law was very loving to the daughter-in-law, dressing her in her best clothes and jewelry. But when the daughter-in-law did not reciprocate her attentions, she threw tantrums, threatening to burn her own hand on the stove or throw herself from the balcony. The marriage broke up, due to the spouses' incompatibility; ironically, the divorced daughter-in-law later discovered her own lesbianism and now lives with a woman partner.

Co-Wives: Till Death Do Us Part?

In a recent collection of lesbian writings from India, a woman called Supriya narrates her story of her conjugal relationship with her co-wife.[32] Supriya was married at 16 to an alcoholic who had no children by his first wife, Lakshmi. Lakshmi had advised her husband to remarry, and she lovingly nurtured Supriya, saying she was like her daughter. She also took care of Supriya's two sons, while Supriya supported the family by working as a domestic servant. The husband could not retain a job due to his alcoholism. Since he visited prostitutes, Supriya was afraid of contracting venereal diseases from him, and Lakshmi protected her from his advances. The two women would sleep together, and their loving friendship developed a sexual dimension. Their husband, whose health was gradually destroyed by drink, was aware of the relationship and tried to turn the women against each other but was unable to do so. Supriya does not know if her children are aware of the relationship but says they respect both women as mothers.

The relationship between co-wives is important and often pivotal in Indian literature. Like the relationship between husband and wife, the relation between co-wives is usually lifelong. Co-wives normally spend much more time with each other than with the husband. Except in cases of extreme conflict, there is always some degree of shared child rearing between co-wives and also between sisters-in-law. Some type of same-sex co-parenting is thus built into the traditional family. This often intersects with adoption within the family. Thus a couple that cannot have children may adopt a nephew or niece, and a woman who cannot have children may adopt a co-wife's child.

Though technically not married to one another, co-wives are marital kin. The Hindi term for a co-mother is *sauteli ma*, often inadequately translated as "stepmother." *Sauteli* derives from *saut*, which means "co-wife."[33] Linguistically and culturally, the function of co-motherhood derives from the status of co-wifehood. Not stepmother but "co-mother" would be a closer translation for *sauteli ma*.

Virginia Woolf remarked that patriarchal narrative highlights hatred between women but ignores love between women. It would be equally true to say that even when narratives depict love between women, readers tend to ignore it and to focus on the more dramatic hatred between women. Thus, the rivalry between Rama's mother and her younger co-wife Kaikeyi is famous because it sets the story in motion, but the loving relationship between Rama's mother and her other co-wife Sumitra attracts far less attention.

The ideal for co-wives as for sisters-in-law (brothers' wives) was to treat one another as sisters, and both co-wives and sisters-in-law were often actually sisters. For example, in the *Mahabharata*, sisters Ambika and Ambalika are both married to king Vichitravirya.

In most premodern societies, women could not maintain friendships and loves in the same way men could, because marriage forced a woman to immerse herself in her husband's life, but did not force a man to immerse himself in his wife's life. In most premodern societies, a woman, after her marriage, moved to her husband's home, and thus lost contact with her family and friends. The sixth-century BC Greek poet Sappho's poems show women friends and lovers lamenting the separation caused by marriage; they miss each other desperately, but are unable to meet. Women in Indian folk songs frequently lament separation from the girlfriends of their youth.

Although in the eleventh-century *Kathasaritsagara*, both male and female same-sex friends are termed *swayamvara* or self-chosen friends, the women's story does not culminate, as that of the men does, with the two friends living and dying together. Somaprabha tells her sworn friend Kalingasena that they can meet frequently as long as Kalingasena is unmarried, but as soon as she marries, they cannot visit each other: "After your marriage, how could I enter the house of your husband? For a friend's husband ought never to be seen or recognized." Thus, while a wife must accept her husband's friend as part of the household, it is considered improper for a man to relate to his wife's friend unless she is a relative. One of the few ways a woman could integrate her woman friend into her family was by making her a co-wife.

In the *Kathasaritsagara*, when the princess Mahallika's marriage is arranged, she persuades her father to get her twelve friends (who are her subordinates, princesses taken captive from other kingdoms) married to her husband. When her husband mildly protests against this group marriage, she scolds him, arguing that since he has already married so many other women, he should have no objection to marrying her friends for her sake. She tells her friends that she would like to have them with her to avoid feeling isolated among her husband's other wives who are strangers to her (VIII. 4. 45). Elevating her friends to the position of co-wives (significantly, they will be co-wives inferior to her in status) will provide her with allies in the husband's household.

A friend tells me that in the early twentieth century, his friend's grandmother got her best friend married to her widowed father. Another woman, who was childless, got her husband married to a girl she herself had raised.

Those married women who lost their girlhood friends after marriage often acquired new friends among their marital kin—sisters-in-law and co-wives. Co-wives could become confidantes, friends, and sometimes lovers. The *Kamasutra* describes co-wives living in women's quarters of royal households dressing as men and using vegetables to engage in intercourse with one another or with their female servants and friends (V.VI.1–2). In the next chapter, I discuss love between women within the women's quarters of the patriarchal household, many of whom were married and some of whom may have been co-wives.

If the husband predeceases them (and, in many cases, being much older, he is likely to predecease them), how do co-wives live out their relationships with one another? The dominant stereotype is that Hindu widows live miserable lives, oppressed and shunned by all. In reality, not all widows live in joint families with men who oppress them. Widows often have greater freedom and mobility than married women, and may acquire the position of powerful matriarchs in the family. The Bengal narratives about the birth of Bhagiratha to two widows, analyzed in chapter 6, provide a glimpse of the more hidden aspects of co-widows' relationships with one another.

Love Between Co-Wives: The Evolution of an Idea

I suggested earlier that medieval narratives often rewrite ancient stories, making explicit what might have been a mere hint or suggestion. One such story is that of Vasavadatta and Padmavati, co-wives of legendary king Udayana. An ancient version of this story glances at the affection between the two women, but a version composed several centuries later develops the relationship into a romantic friendship based on attraction, which works as an alliance to keep the husband in check.

In *Swapnavasavadattam* (Dream of Vasavadatta), a Sanskrit play by dramatist Bhasa (ca. 275–335), queen Vasavadatta, wife of King Udayana, is compelled to assume a disguise, and meets the young princess Padmavati by accident. The two women are intuitively drawn to one another. They notice each other's beauty, and, within a short time, each says to herself that the other has become her own (*atmiya*). Later, Vasavadatta lives as Padmavati's close companion and friend. King Udayana thinks Vasavadatta is dead and agrees to marry Padmavati. Vasavadatta is reluctantly compelled to weave Padmavati's wedding garland.

A hint of the closeness between the two women appears when, soon after the marriage, Padmavati develops a headache (perhaps partly due to having heard her new husband bemoaning the loss of Vasavadatta). Vasavadatta goes to tend Padmavati, and thinks she sees her lying covered and asleep on the bed (in fact, it is the king lying asleep there). Vasavadatta remarks, "Perhaps by keeping a part of the bed empty, she is indicating, 'Embrace me.' So I will lie down."[34] It is interesting that although Padmavati is newly wed, Vasavadatta thinks that Padmavati wishes to be embraced in bed by her woman friend. This indicates that the two women have probably been bedfellows before.

About eight centuries later, the *Kathasaritsagara* fleshes out the two women's relationship. As in the earlier version, the king's chief minister decides it is politically expeditious for the king to marry Padmavati, princess of Magadha. However, he knows Udayana will not consent because he is in love with his wife Vasavadatta. The minister therefore persuades Vasavadatta to go away in disguise and he burns her palace, so that her husband thinks she is dead. In the earlier version, Vasavadatta was not aware of the minister's plan to remarry her husband; in the later version, she is aware of it, and therefore decides to check out her prospective rival.

She goes to Magadha in disguise and meets princess Padmavati, who she thinks is destined to displace her, but here the story takes an interesting twist: "And Padmavati, when she saw the queen Vasavadatta . . . fell in love with her at first sight" (II. 16) (*Devim Vasavadattam tam drishoha pritirajayata*).[35] The description of Padmavati's attraction to Vasavadatta is not merely friendly but sensuous: "And Padmavati perceived that Vasavadatta was a person of very high rank, by her shape, her delicate softness, the graceful manner in which she sat down, and ate, and also by the smell of her body, which was fragrant as the blue lotus" (I. XVI).

The love and friendship that develops between the two women makes co-wifehood acceptable to them. Vasavadatta signals this by making unfading garlands and *tilaks*, symbols of marriage, for Padmavati. Vasavadatta's gift of these items to Padmavati signals her acceptance of her as co-wife prior to the wedding ceremony. The female-female bond precedes the female-male bond. Vasavadatta repeats this action just before the wedding, again making garlands and *tilaks* for Padmavati. Appropriately, it is these garlands and *tilaks* that reunite Vasavadatta with her husband, because their exceptional quality leads him to ask his new bride who made them, whereupon Vasavadatta is revealed.

The two women suffer depression and jealousy once they are established as co-wives; however, they rapidly overcome these feelings and declare that they are sisters. Their feeling of oneness is cemented when they learn that they were "sister goddesses in a former birth" (I. XVII). Vasavadatta "made her husband equally the property of both" (I. XVII), and Padmavati's father, who was angry when he heard his son-in-law already had a wife, is appeased when he realizes that the two women have "but one heart" (*ekam hi hridayam*) (I.XVI.169). Later in the story, when the king wants to marry yet another woman, Vasavadatta and Padmavati band together and successfully pressure the minister into preventing this third marriage.

The relationship between sisters-in-law (brothers' wives or a man's sister and his wife in the joint family) is in some ways similar to the relationship between co-wives. The norm is that they should love one another as sisters, the stereotype is that they are jealous rivals, and the reality as well as the textual representation are often more complex than either ideal or stereotype.

Sisters-in-Law

The domestic space, idealized in modern Christendom and Hinduism as the domain over which a wife reigns as queen of the family, sometimes becomes hospitable to female-female liaisons. Several works of twentieth-century Indian fiction represent female friendships developing into love relationships, within the marital home of one or both women. In Shakespeare's *As You Like It* (1598), Celia, who vows undying love for her cousin Rosalind, and accompanies her into exile, ends up marrying Rosalind's husband's brother, which results in their becoming sisters-in-law.[36]

In the 1998 Hindi film *Fire*, two brothers' wives living in a joint family become lovers. Occasionally female lovers marry two brothers so as to continue their relationship within the respectability of a sister-in-law relationship. An Indian lesbian, now living in the United States, recounts how, in her first teenage relationship in India, she suggested to her lover, "we should find a pair of brothers to marry so that we could live in the same house and continue our relationship. It seemed the closest thing to what we viewed as normal." Another option they considered was suicide: "several times . . . we talked about killing ourselves . . . because the world around us was so potentially hostile that at times death seemed like the only way out."[37]

In another case, in June 1997, police arrested a teenaged apparently heterosexual couple in Ghaziabad, an industrial town in north India. The neighbors suspected the two of being minors who had eloped together, and reported them to the police. When the police searched the boy, Mukesh, they found that he was a girl, Sita, who had eloped with her lover Pooja from their homes in nearby Ambala six months earlier. The girls did not want to return home nor did their parents want them back. The parents said the girls had eloped before, whereupon the parents had suggested that they marry two brothers "which would ensure that they live in the same house."[38] Although discreetly stated, implicit here is the idea that they continue their relationship within the safe boundaries of heterosexual marriage. The girls, though, had refused the suggestion.

"Who Can Speak of Men?" (2003), a documentary film made by Ambarein Qadar, Ghazala Yasmin, and Nihal Waqfi, students of Mass Communications Research Center, records the stories of two female companions, Kafila and Nigar, one of whom wears male attire. At one point the women say that if they have to marry, they would like to marry two brothers and stay in the same house. They add laughingly that if they do this, their husbands may not be happy because the two women would always be together.

Why Not Stay on the Down Low?

In the current debate in India about homosexuality, several commentators have suggested that since homosexuals are not extensively persecuted or attacked in India, and since many manage to maintain secret same-sex liaisons within the conventional family, there is no reason to construct sexual identities and demand minority rights, along the lines of LGBT movements in the West. Similar arguments appear in the West in slightly different form. Except for some religious extremists in the United States who believe that homosexuality can be eradicated, average heterosexists prefer that homosexuals keep their lives private and out of the public domain. Even liberals often express discomfort with same-sex marriage, suggesting that what happens in the bedroom is private and should not be "in your face" in public spaces. Many families, both Indian and American, respond to a child's coming out, by suggesting that s/he be discreet and not embarrass the family with public statements or actions, such as same-sex marriage.

At the opposite end of the spectrum, many left-wing, gay-friendly HIV and AIDS activists, both in India and the United States, insist that men married to women who have relationships with men on the side should not be morally judged for deceiving their wives or identified as homosexual or bisexual. Rather, they should be supported in their choice to live as heterosexual married men while engaging in same-sex relations on the side. This view is problematic not only because of the subordination of women on which it is premised (few of these men would be happy about their wives engaging in extramarital affairs) but also because of its assumption that supporting the status quo is preferable to making more choices available.

While demonstrating that same-sex desire has in the past existed and still does exist within traditional families, I do not mean to suggest it flourishes there. Among the gay Indians I know who have entered heterosexual marriage without telling their spouses, almost all have been plagued by fear, guilt, shame, or regret.

In her autobiographical novel, *Madras on Rainy Days* (2004), an Indian Muslim woman, Samina Ali, recounts how her gay husband was unable to consummate their marriage. Mathematician Shakuntala Devi (born 1939) describes how her husband's gayness led her to write her pioneering book on the status of gay people in India.[39] The few exceptions are those where both spouses are bisexual, or one is heterosexual and the other gay or bisexual, but they reach a mutual agreement not to be monogamous.[40] I do not have the data to examine the relative happiness of MOCs.[41]

Even in the few cases where a gay person is relatively happy in a heterosexual marriage, s/he often looks back with nostalgic longing to the same-sex love ended by that marriage, as Vita Sackville-West does in her diary. This landscape, riddled with grief, loss, and pain, is poignantly represented in Indian fiction. In her pioneering Malayalam story, *The Sandal Trees*, Kamala Das shows two young women lovers separated by marriage to men. At the end of the story, the elderly protagonist realizes that she has always shortchanged her husband because she has never forgotten her girlhood beloved.[42] In Shobhana Siddique's Hindi story *Lab-ba-Lab* (Full to the Brim), a young married woman remembers with longing her premarital relationship with another woman, even as she adorns herself to spend time with her husband's aunt whom she finds attractive.[43]

Fiction also indicates how some married people use heterosexual privilege to exploit and discard same-sex lovers. The "safe" same-sex lover, like the "safe" cross-sex lover, is often a social subordinate. From Ismat Chughtai's "*Lihaf*" (Quilt, 1942) to Anita Nair's *Ladies Coupe*, married people are shown being pleasured by same-sex servants. Cross-sex sexual relations with servants are also widely represented in literature. In Nair's novel, both husband and wife use the maid, Mari, sexually, but Mari is in love with the wife, of whom she says, "The first time I saw Sujata Akka, I lost my heart to her."[44] Sujata finds sex with her husband disgusting, and invites Mari's advances but never reciprocates. Mari doesn't mind, saying, "I had loved her with my heart for so long; it seemed natural that I love her now with my body"(261). But when Sujata discovers that her husband has also been having sex with Mari, she throws Mari out, dubbing her "unnatural."

Heterosexually married people who engage in same-sex relations often cling to marriage not just for social privilege but for psychological protection from the stigma of being "unnatural." Thus, in cross-class relations, it is not always or only the subordinate who is exploited. When the social superior is single and the subordinate married, heterosexual privilege may trump class privilege and result in a transaction that is mutually useful but that abjects the gay person. This happens in R. Raj Rao's *The Boyfriend* (2003), where the gay protagonist, Yudi, a journalist, manages to retain his relationship with his working-class boyfriend Milind only by supporting him financially, while Milind, who gets married to a woman, despises Yudi for his homosexual identity and single status.

From the vantage point of the married person, same-sex intimacy, though sought for sexual and emotional satisfaction, is often episodic. This pattern is depicted in Manju Kapur's *A Married Woman* (2003), where the protagonist Aastha has an intense affair with a widow turned lesbian, but finally returns to her colorless but not entirely unhappy marriage.

Other Families

If the mainstream Indian family, nuclear and extended, is not as conventional as it appears on the surface, it is also important to remember that openly unconventional family arrangements also exist in India.

Courtesan lineages, now rapidly dying out, constitute a type of family arrangement wherein female bonding has a more important institutional function than it does in the mainstream patriarchal family. Different types of courtesan families, ranging from *devadasis*, discussed earlier, to *tawaifs*, who were singing and dancing women patronized by Muslim nobility and gentry, were organized more around women, who were the primary earners, than around men. Daughters were the professional and financial heirs of their mothers, while sons were trained to function as assistants of various types.

In the next chapter, I discuss the representation of female-female sexual relations in two all-women spaces—the private women's quarters of the patriarchal household and the public space of the courtesan household.

Chapter 8

"Married Among Their Companions": Female-Female Relations in Premodern Erotica

You and I sitting here like bride and groom,
Let's agree on a dower of a lakh rupees, O dogana.
—*Insha*

Before the twentieth century, explicit depiction of same-sex sexual relations was found mainly in erotic literature, much of it written by men. Shunned by Victorian litterateurs as obscene and dismissed by many feminists as degrading to women, these depictions have not until recently received serious attention. While gay male historians often acknowledge and draw on these writings, lesbian historians, influenced by feminism, tend to characterize them negatively.[1]

Indian literatures' many traditions of writing about the erotic fell under a shadow in the nineteenth century, as did erotica in Euro-America. In this chapter, I examine the depiction of female-female relations in a genre of nineteenth-century erotic Urdu poetry called *rekhti*. *Rekhti* depicts female-female relationships as institutionalized in various ways, including in marriage-like unions. These poems are little known, and several are here translated into English for the first time.

Rekhti developed from the late eighteenth century onward under the Indian Islamicate.[2] All of the major *rekhti* poets were Muslim, and the poems depict women living in Indo-Muslim households.

Rekhta (literally, scattered) is another name for Urdu, a language that evolved in the medieval period on the Indian subcontinent, from a mixture of Persian and Sanskrit-based languages.[3] *Rekhti*, the feminine of *rekhta*, is a term coined by poet Sa'adat Yar Khan (1755–1835), whose pen name was "Rangeen" for the genre of Urdu poetry that emerged in southern India in the eighteenth century, and of which Rangeen himself was a prominent exponent.[4] Most *rekhti* poets were men. The work of female *rekhti* poets mentioned in the sources has unfortunately all but vanished.[5]

From the mid-nineteenth century onward, *rekhti* poetry, along with Urdu poetry in general, and much literature in other Indian languages, was "reformed" and "purified" by literary critics, editors, and poets, as a result of which *rekhti* grew more didactic and less erotic. In the course of the twentieth century, *rekhti* poems by earlier poets were censored by editors or became unavailable to readers.[6]

The three major poets whose verse depicts female homoerotic relations are the aforementioned "Rangeen" (Colorful), Insha Allah Khan (1756–1817), whose pen name was "Insha" (Elegantly Stylish), and Shaikh Qalandar Baksh (1748–1810), whose pen name was "Jur'at" (Audacity). In this chapter, I refer to the poets by their pen names. I also cite the work of minor *rekhti* poets whose work is even less accessible than that of the major poets.

The Language and Rituals of Female Intimacy

Rekhti represents love between women with complexity and in several registers. The tone is often humorous, even camp. *Rekhti* poems sometimes explicitly describe sexual intimacy between female lovers; more often, the speaker praises her lover's beauty and expresses feelings of love and fulfillment or longing for union and anguish at separation. There is considerable concern about concealing the relationship, to preserve the women's reputation.

Rekhti uses several terms to refer to women with a predilection for sexual intimacy with other women, and also to their sexual activities. These words are found in other contemporaneous sources as well, and, although not widely used today, some were still used in certain contexts in the late twentieth century.

Perhaps the most common are *chapat* or *chapti* or *chapat bazi*, all words for female homosexual activity; the women engaging in such activity are called *chapat baz*. The suffix *baz* indicates an agent, player, or fancier, as in *shatranj baz* (chess player) or *kabutar baz* (pigeon fancier). In his 1884

Urdu-English dictionary, John Platts, defining this word, retreats into Latin in the approved fashion of Victorian lexicographers: "*capatbaz, s.f. Femina libidini sapphicae indulgens;—capatbazi, s.f. Congressus libidinosus duarum mulierum.*"[7] He provides the same type of definition for other usages of the term, such as *Chapti khelna*, *chapti larana*, and *chapti larna* (all signify "to play *chapti*").

Chipatna (in Urdu spelled the same way as *chapatna*) means "to stick, to adhere; to cohere; to cling to."[8] A similar notion of lesbian sex is contained in the French term *tribade* (also used in English sixteenth-century onward) from Latin *tribas* and Greek *tribein* (rubbing), as well as the Arabic term for such activity, *sahq* (rubbing). Similar too is the term *fricatrice* (from frig, to rub, Middle English *friggen*). In one poem that I analyze later in this chapter, the Urdu word *fe'l*, which connotes homosexual acts in general, is also used to refer to sex between women.

In her interviews in the 1970s with courtesans in Lucknow, a city in north India renowned as the capital of the Islamic kingdom of Awadh, Veena Oldenburg notes of those who had relationships with each other, "They referred to themselves as *chapat baz* or lesbians, and to *chapti*, or *chipti*, or *chapat bazi* as lesbianism."[9] This is very important, because it shows a continuous history of the word, at least among Lucknow courtesans, who figure largely in *rekhti*, and from whom poet Rangeen claimed to have learnt the idiom he used in *rekhti*.

Another nineteenth-century source testifies to the use of these terms and to awareness of female homosexual relationships. In 1900, British sexologist Havelock Ellis quoted a report sent to him by an officer of the Indian Medical Service, which cited five words for homosexually inclined women—*dugana*, *zanakhi*, *sa'tar*, *chapathai*, and *chapat baz*. The officer said that two women who lived together were referred to as "living apart," and gave examples of such women he had come across, including an intercaste couple living in a town, a widow who had relations with her three maidservants, a couple in prison, and a pair of widowed sisters. The officer also mentioned *rekhti* poetry: "The act itself is called *chapat* or *chapti*, and the Hindustani poets, Nazir, Rangeen, Jan Saheb, treat of Lesbian love very extensively and sometimes very crudely."[10]

Chapti and its variants occur in *rekhti* poetry, as part of a cluster of terms that indicate special intimacy and erotic activities between women. These terms overlap with others that indicate fictive kinship. The most frequently used term is *dogana*. According to Platts' dictionary, *dogana*, or *dugana*, from the root *do* meaning "two" (Pehlavi "*du*," Old Persian "*doa*," Sanskrit "*dva*") has several meanings, among which are "Double, two together" and "two intimate friends, an inseparable pair."[11]

This term is still used in Urdu today for a fruit that has two bodies in one shell, for example, when what look like two mangoes or two bananas are enclosed in one skin. Rangeen, in his definition of *dogana*, cites this meaning (of a doubled fruit) as ritually related to the institutionalization of the *dogana* relationship between women.

In *rekhti* poetry, the term *dogana* is used both to refer to and to address a woman's intimate companion. It is also used to refer to the sexual intimacy between them: "I am given to *dogana*/ as the moss is given to greenness."[12] In some poems, *doganas* are referred to in the plural, as women who engage in female-female sexual intimacy. When used to refer to a woman, *dogana* thus becomes a term that indicates a sexual predilection. Other terms used to refer to or address a woman's intimate female companion are *zanakhi* and *ilaichi* (literally, cardamom).

Here is a summary of the rituals whereby a long-term relationship between two women was established:

> *Ilaichi* and *dogana* and *zanakhi* and *dost* and *sahgana* and *guiyan*, all these have the same meaning. Those women are called "*ilaichi*" [cardamom] who in private together feed each other the grains of a cardamom, and those are called "*dogana*" who each eat one of twin almonds and become *dogana*, and those call each other "*zanakhi*" who take a chicken wishbone and break it and become *zanakhi* together, and moreover they customarily call that one "*sihgana*" who is the *dogana* of a *dogana*, therefore there is extreme jealousy towards her but for the sake of the *dogana* or to tease and create a tempest she is called a *sihgana*. The purport of all this is that these relationships usually exist mutually among those women who engage in *chapti*.[13]

The term *sihgana* (literally, "of three kinds,") sets up a triangulation between a woman, her *dogana*, and the *dogana*'s *dogana*. The only use of the term in *rekhti* I have come across is in a verse by the poet "Begam" (literally, Lady):

> My *dogana* went as a guest to the *sihgana*'s house,
> I rolled on hot coals, my life left me. (IR 83)

To roll on hot coals is a popular idiom indicating emotional anguish. There is a play on the word "life"—since a beloved person is often addressed as "*Meri Jan*" (my life), the "life" that left the speaker could be her beloved *dogana* while the phrase also suggests that her anguish at her rival's victory threatened her life. This type of hyperbole is typical for the lover in the standard Urdu *ghazal*.

In the glossary to his *rekhti* collection, poet Rangeen also provides a more detailed account of the rituals involved in setting up the relationships of *dogana*, *zanakhi*, and *ilaichi*:

> They [feminine] send for almonds from the market and take out their kernels. When a twin or *dogana* almond emerges, one kernel is always embedded in the other. Calling the one that is embedded male (*nar*) and the one in which it is embedded female (*mada*), they call an unknown person, give him both kernels and tell him to give one to each of them. The one who gets the male kernel considers herself the man (*marad*) and the one who gets the female kernel is compelled to (*majbooran*) become the woman, and both mutually are known as *dogana*.[14]
>
> *Zanakhi*: They [feminine] slaughter a chicken, get it cooked, and then sit down to eat together. In the chicken's breast is a divided, double-branched bone [wishbone] that is called *zanakh*. Each holds one end of the bone and pulls it toward herself. The one whose end breaks off is called the female and the one whose branch remains is called the male. And if it snaps in equal pieces, they get another chicken slaughtered and repeat the process until the identification of male and female is completed.[15]

All three rituals involve eating together. Rangeen also describes all three types of relationships as establishing "male" and "female" roles between the two women.

"Then They Get Married"

The most interesting element of the *ilaichi* relationship as described by Rangeen is the reference to two women marrying each other, which, to my knowledge has not so far been translated or commented upon:

> *Ilaichi* [literally, cardamom]: Two women each take a cardamom and break it open. The one in whose cardamom there is an even number of seeds becomes the male and the one in whose cardamom there is an odd number of seeds is compelled to become the female. If both get identical orders of seeds, they repeat the ritual until odd and even numbers emerge. Then they get married together among their [female] companions, and these are called ilaichi.[16]

That the marriage occurs in an all-women context suggests that the two would be recognized as a couple only among their women friends. These marriages between women are different from most male-female marriages insofar as they are not public, and are thus accorded a lower status, and also insofar as they are founded upon mutual attraction and choice. In the latter respect, they are similar to the unions between male friends mentioned in the *Kamasutra* (see chapter 1).

Eating together or feeding each other is a central ritual in establishing relationships in India. It is part of many wedding ceremonies. Cardamom is eaten to sweeten the breath and is associated with love. In the film *Fire* by Deepa Mehta (1998), the sisters-in-law, after they become lovers, feed each other cardamom, and teasingly discuss the fact that brides are asked to eat cardamom.

Like Bride and Groom

The idea that two women in love may marry one another, and that one of them would then be the wife and the other the husband, is found as early as the second century AD in West Asia. The satirist Lucian was a Syrian who grew up on the banks of the Euphrates. The Roman Empire at that time stretched from what is modern Iraq to what is modern England. Lucian made his way to Athens, and became a Greek scholar. Lucian's *Dialogues of the Courtesans* contains one of the earliest depictions of female-female sexual relations. Courtesan Leaena tells her friend Clonarium about her sexual relationship with Philippa, a wealthy woman from Lesbos, who woos her with expensive gifts and seductive caresses. Philippa, a manly woman, matter-of-factly states: "Years ago I married Demonassa here; she's my wife."[17] Demonassa also participates in the seduction of Leaena, and she is in bed with Leaena and Philippa when their relationship is consummated.

Rekhti poetry too depicts marriage-like unions between women, some of them courtesans. In one of Insha's verses, the woman speaker declares:

> Yun hi main gash hui dogana par
> Raja Nal jaise tha Daman par gash.
> (I swooned [with love] over my *dogana*,
> As King Nala swooned over Damayanti.)[18] (KI 416: 43)

Just as modern newspaper reports compare female couples who commit joint suicide to Romeo and Juliet, the lovers here compare themselves to legendary spouses Nala and Damayanti, famed for their fidelity. This suggests that the female lovers aspire to a conjugal ideal. Since, in popular imagination, only male-female relations are sexual, while female-female relations are devoid of sexuality, the speaker's comparison of her feelings for her lover to Nala's feelings for Damayanti also sexualizes the female-female relationship.

Other verses too emphasize the desire for long-term commitment and public union. In a couplet by Rangeen, the female speaker cries: "O God, may no one be inclined to desire,/And if they are, may they be inclined to commitment" (IR 13:16). In a couplet by Insha, she contentedly remarks to her lover:

> You and I sitting here like bride and groom,
> Let's agree on a dower of a lakh rupees, O *dogana*. (IR 22:5).

The word translated as "dower" is "*mehr*," the money a Muslim groom either gives his bride's father at marriage, or promises, in the marriage contract, to settle on her. The terms "bride" and "groom" invoke the cultural reality of marriage. This invocation occurs too in the reports of weddings between women in India from the 1980s onward.

My Sister, My Spouse

That a *dogana* may be a lover and spouse does not preclude her also being a sisterly figure. As discussed in chapter 1, fictive kinship relations are not equated with biological relations, and incest taboos do not affect them in the same way. A fictive sibling relation between man and woman sometimes paves the way for a romantic relationship. This dynamic is also represented in Hindi movies, in popular Urdu nineteenth and twentieth-century fiction, and in nineteenth-century English fiction, in which the hero frequently proves his reliability by establishing a fraternal relation with the heroine, and is even termed her brother, before he proposes marriage in the last chapters.

Similarly, fictive sisterhood can coexist with a lover relationship between women. Fiction in several Indian languages, by both men and women, shows female romantic friends and lovers addressing one another as sister—"*chechi*" (Malayalam), "*didi*" (Hindi) or *apa/baji* (Urdu).[19] In some of these texts, when the fictive sisters are lovers, one woman is shown behaving in a more masculine and the other in a more feminine way. Thus, the fact that the female speaker in *rekhti* may sometimes address her *dogana* as "sister" does not mean that the relationship is necessarily non-amorous.

The Ups and Downs of Love

Female-female love in *rekhti* goes through all the ups and downs that male-female love and male-male love conventionally go through in Urdu poetry. The female speaker frequently expresses appreciation of her *dogana*'s beauty: "Why should my heart not throb in my breast? [literally, life throb in my liver] /Your beauty is like gold" (KI 403: 15). Sometimes, she gives a head-to-foot description (*sarapa*) of her lover's attractions. This literary device, known as the blazon in medieval European poetry, is known as *keshadipadavarnana* in Sanskrit and *nakh-shikh-varnan* in Hindi. Each feature of the beloved (lips, teeth, breasts) is praised as the best possible. Rangeen's speaker gives the convention a twist by insisting that the *dogana*'s features are not typical but unique: "Ah, my *dogana*'s style is quite unique—/She's cream-complexioned with a special magnetism . . . Her way of talking is different from all others', every detail of her appearance is unique. . . . Everything about her is different from everyone else/She goes at her own unique pace" (IR 14: 22).[20]

Dressing up together is a form of bonding: "These days of the new year are green/ Buy green clothes for me, put them on me, come" (Rangeen 11: 4).[21] Teasing references to undressing also occur: "O fairy, what can be said of your drawstring? / Your drawstring is the loveliest of all drawstrings/ Lightning seemed to flash before my eyes/—Your drawstring spread through the dark clouds" (KI 411: 31). The editor has excised four couplets following this tantalizing one.

But sexual attraction is not always based on dress:

It's not the way your jewels adorn you—
I am enamored of your simplicity.
Come, O Rabeel [woman's name], today
I am dazzled by your fair, fair body.

. . . .

The breeze of silence blew
My heart was stunned by its sound.
Walking in the garden, "Insha,"
I swooned over the garden of your pajamas. (KI 416: 43)

The female speaker is often anguished by separation: "My heart feels paralyzed, *dogana*, at that moment/When you call for a palanquin to take you home" (KI 428: 69). This verse indicates a situation where the women live in different households.

The *dogana* is sometimes inattentive like the cruel beloved in the standard Urdu *ghazal* or like Krishna in Hindi love poetry:

Oh heart, she takes no account of you
Your wretched desire has no effect.
Why do I not complain and lament?
Because you give no thought to my condition.

If she ever does say "yes" one moment,
She follows it up by saying "No" for two watches more,
When I said, I am fainting, that fairy
Replied, "Don't worry, you won't."
I would like to fly away as the wild ducks do.
Alas, Insha, I have no wings. (KI 425: 61)

The speaker in *rekhti* reproaches her *dogana*:

So you turned out false once more, O unjust one.
Yesterday too you broke your promise to come, O unjust one.
Look, you are like a lamp's flame emerging from my eyes,
You burn my heart, O unjust one. (KI 419: 51)

When they quarrel, the speaker complains: "How can I describe my *dogana*'s indifference?/All willfulness is based in her mind alone" (KI 398: 4), and may even declare herself relieved when they break up: "My friendship with the *zanakhi* ended, that's good/ Nurse, the fetters have been cut off my feet" (KI 399: 6).

Letters, Trysts, and Social Stigma

Pleasure in the *dogana*'s passionate love is accompanied by fear of being discovered:

I keep my *dogana*'s letter in my bodice
This paper has become like a boil on my breast.
Tell her not to keep writing and sending me letters all the time
This wretched paper will defame me. (KI 412: 35)

The speaker expresses dismay at her *dogana*'s open display of their relationship, which potentially exposes them to social disgrace and perhaps to unwanted advances from others, such as the young letter-reader and carrier, teasingly identified with the poet-persona:

You have worn my scarf, *dogana*, this is awkward
This stain will sully both of us, *dogana*, this is awkward.
Don't behave thus and chuckle delightedly—people will say
That you and I are having an affair, *dogana*, this is awkward.
You are cold and dry, bitter and acerbic to me by turns,
Whatever the sweet and sour, *dogana*, this display is awkward.

This tussling is not a good idea, let such things go—
There is a nail-mark on my face, *dogana*, this is awkward.
I need an old man on the doorstep to read letters—
Insha is young and strong, *dogana*, this is awkward. (KI 403–404: 16)

Until recently, most families in urban India slept on rooftops on summer nights. Women and men of the household slept on separate parts of the roof. This provided little privacy to spouses or lovers, and fiction often depicts romantic situations that ensued. In *rekhti*, the speaker and her *dogana* visit one another by night on the rooftop, which is both exciting and dangerous. Sometimes, both women belong to the same household and the danger is that the sounds of their lovemaking may awaken others. At other times, they climb on to one another's rooftops, and creaking doors may betray them. Meeting in the day is also difficult because other women are around. These problems are variations on premodern poetry's general theme of love's danger:

Lady, love is a digression
That has uprooted thousands of homes. . . .
At night I scaled your rooftop with a ladder
And hid behind the parapet.
I wish your hinge would break,
You wretched, unmelodious door.
Why do you bring along with you
These troops upon troops of young girls?
It can't be helped—I know that in love
There are thousands of ups and downs.
As long as you can, O *dogana*, my life!
Keep trying your best—
Ultimately, fate or destiny will decide
Whether we survive or are ruined. (KI 413–414: 38)

In another poem, the speaker, addressing an unidentified person, vehemently denies that her lover visited her at night:

When did *zanakhi* come to me last night—that is false.
How could she and I meet in any wrong way?
My string-cot was laid there—by which route could she have come,
Climbing over such a high wall! This is totally untrue.
Is she a bird that could fly and reach here? . . .
Apart from love's attachment, all other magic is false. (KI 417–418: 47)

The Debate on Female Homosexuality

Rekhti represents characters with different points-of-view on many subjects, including female-female love. Some speakers criticize and denounce such love; others celebrate it; others complain of it; and yet others start off wary and critical, but warm to the idea by the end of the poem.

This debate is part of a larger debate on love in general, which, in premodern literature, is generally seen as a force that threatens social institutions such as marriage. In medieval romances in Indian languages, the heroine's friend frequently warns her of the dangers of love, a warning the heroine usually disregards. *Rekhti* participates in this convention. Here is a typical example:

> Ladies, think of the breadth of the river of love
> Do not fearlessly set foot in it, first think of your home
> Mountains are washed away in it, what of a boat?
> Love's edge is sharper than a sword, think how it may cut you. (KI 408: 25)

In some *rekhti* poems, the female speaker struggles to refuse offers of illicit love, both heterosexual and homosexual:

> May no one be defamed by love of any person.
> O Nurse dear, may such bad acts never be committed.
> If the *doganas* get annoyed with me, let them.
> May I never have the gift of union with them.
> That man said to me, "Come, let's rest awhile,"
> May I never have such "rest" as he calls rest. . . .
> As long as it's daytime, my heart is safe, Insha,
> May that inauspicious terrible black evening not come again. (KI 407: 23)

In another poem, the speaker expresses shock at a woman's attempt to seduce her:

> Sister, the kind of friendship you want from me
> Oh, may such friendship not exist between two unmarried girls!
> Don't try to talk me into it, get lost
> What is it you are calling love, what kind of friendship is this? (KI 416: 44)

However, this type of speaker is rare in *rekhti*. Significantly, she emphasizes that both she and her would-be seducer are unmarried girls or virgins (*kunvari*), who therefore should not engage in sex of any type. We may compare the *Manusmriti*'s anxiety about a girl losing her virginity through lesbian activity (see chapters 1 and 6).

The debate on female-female sexual activity is similar to medieval poetic debates on other topics, such as the nature of women. In the seventeenth-century Punjabi romance *Heer Ranjha* by Waris Shah, the hero Ranjha engages in a spirited debate with his beloved's sister-in-law Sahiti. Ranjha denounces women while Sahiti defends them and denounces men.[22] These debates usually remain unresolved, with each participant remaining convinced of his or her own point of view. The same is true of the debate in *rekhti* on female homosexual relations.

A Dream of Defiance

Two poems by Jur'at and two by Rangeen are titled "*Chaptinamas*" (*Chapti* Narratives), and focus exclusively on female-female liaisons.

One of Jur'at's *Chaptinamas* tells the story of two women named Sukkho and Mukkho. The refrain, repeated after every quatrain, runs: *Ao dogana, chapti khelein, baithe se begar bhali* (Come *dogana*, let us play at *chapti*, better to labor without payment than to sit idle").[23]

Both Sukkho and Mukkho are married and appear to live in the same household. The poem begins with them declaring that the heart's bud can blossom only if one wanders through the garden, and that staying with their wretched husbands has made their lives miserable. Therefore, they decide to invite over all the *chapti baz* women in town, and entertain them with flowers and betel. They are aware that such behavior is socially censured, but decide to take on the opposition. When older women lecture them, they laughingly exchange private gestures. The pleasure of their love is so great that they do not care if the whole household grows hostile to them (*go dushman sara ghar ho*).

Sukkho and Mukkho state that the pleasure of love between women "is better than all others in the universe":

> To the enjoyment of *chapti* what other pleasure can compare?
> This rubbing above, below, is intercourse wondrously rare.
> Making love with one's likeness is a strange, delightful thing.
> Even if you get entrapped, being so consumed is comforting.(*SSLI* 222)

The symbol of wine, which represents love's intoxication in the standard *ghazal*, appears here. The difference is that the male lover in the *ghazal* drinks wine openly while these female lovers drink wine in secret: "*Chori*

chori peejiye katey nashey mein khush auqat" (Drink this wine in secret, get drunk, the time will fly happily). Finally, a female servant betrays them to "the Mirza," and they fear that he will now imprison them at home. In the last stanza, the poet remarks:

Sukkho aur Mukkho ke jur'aat-e fe'l karun kya aur insha
Jab donon ke khasmon ne sab baton se un ko mana kiya
To kahti thin chapatbaaz to hain mashhoor ab hum har ja
Kyon na amal phir is par kijey nachney nikley to ghunghat kaisa

(What else can I elegantly write about Sukkho and Mukkho's daring intercourse? When their husbands forbade them to do what they were doing They said, "We are now famous everywhere as *chapat baz* Why not act upon it then—when going out to dance, why wear a veil?") (*SSLI* 223)

The word for the women's "daring" (*jur'aat*) is a play on the poet's pen-name Jur'at and gestures toward his own daring in writing on such themes. "*Jur'aat*" is interestingly combined here with "*fe'l*," a word which, according to Platts, signifies "A deed, an action, act; labour, work" and also "carnal intercourse" but can also refer to adultery, prostitution, and "an unnatural offence" or "an unnatural act," the latter being the standard Victorianism for homosexual acts.[24] Jur'at thus categorizes female intercourse under homosexuality in general, and also includes a play on fellow-poet Insha's pen-name (Elegantly Stylish) by way of praising his own writing.

The refrain of Jur'at's second *Chaptinama* is: "The way you rub me, ah! It drives my heart wild/Stroke me a little more, my sweet *dogana*" (*Ragda de zari aur meri pyari dogana*) (*SSLI* 223). The speaker expresses romantic feelings: "I'd give up anything for that moment when you come in,/ Dressed pretty as a picture, and put your arms around my neck" (*SSLI* 224) and also a clear sexual preference for a woman over a man: "Let her go to men who wants stakes hammered into her—/ Can she ever get these hours and hours of pleasure?"(*SSLI* 224). A repeated point of comparison is the extended caressing of a woman as compared to quick intercourse with a man: "That wretched man should feel ashamed of coming so soon" (*SSLI* 225).

Various sexual activities between women are referred to, including mutual rubbing of vulvas, stroking and rubbing with fingers, and the use of dildos. A play on the word "tongue" (meaning, as in English, both the bodily organ and "language"), which is popular in Urdu verse, appears: "When I take your tongue in my mouth and suck on it—/ With what tongue shall I describe the state I am in?" (*SSLI* 224) The speaker praises her *dogana* as far more skilful than all the other women who engage in *chapti*. The secrecy of the relationship heightens its pleasure: "How to describe the taste of sweets eaten in secret?" (*SSLI* 225).

This poem is remarkable for its generalizations about the pleasures of love between women: "When one woman clings to another, such is the happiness/They never want to part or let their desire decrease" and "There's no pleasure in the world like clinging to a woman."(*SSLI* 225) The poem concludes with an emphasis on choice and romantic intimacy rather than lust alone:

> However much "daring" a man may have
> However much energy and lustful desire
> I'd rather see a face that gives me pleasure—
> I'd give anything for this intimacy, which I much prefer. (*SSLI* 225)

The word "daring" is a play on the poet's name, Jur'at. The poet-persona thus playfully brings himself into the poem, and acknowledges that in the speaker's eyes his own sexual charms pale before those of her female lover.

The Pleasures of Sex

In Rangeen's poems, as distinct from Jur'at's, the emphasis is much more on sex than romantic intimacy. Here is his first *Chaptinama*, here translated into English for the first time:

> Sister, I went to the park yesterday
> What a sight I saw there!
> In one area there, I saw
> Two whores under a tree.
> One moved on top of the other
> This is the way one united with the other
> Sometimes one would kiss the other's mouth
> This is the way she would press her breast
> Sometimes one would tickle the other
> And let her hair fall upon her cheek
> She would express her readiness to die for her
> And the other would give her many kisses.
> When one pushed the other hard
> She would cry, Oh I am dying!
> What coquetry there was in their gestures
> And in all their blandishments.
> As they united, I was abashed.
> One would tease and excite the other.

When she got hot and began to writhe
The other would slow down her movements.
The one below would say,
"Apply your arms and sometimes your shoulders
Wipe your sweat with your kerchief,
Don't spoil the pleasure of this condition.
Rub lip to lip, my sweet,
I am dying of thirst, give me some water.
Do rub your breasts against mine
So that your nipples are not seen.
Wrap me in your arms and squeeze me tight
When I writhe, catch hold of me
You are twice as strong as I am -
How come you have grown cold?
Dogana, whenever you quench my excitement,
I become forever your slave girl.
If you can't, then get below me and let me be on top
Follow my example, watch how I unite with you."
When they both came together
They began to say to one another,
"One who has a dildo
Only such a one has the whole game."
The one below said, "O *dogana*, my life,
I would sacrifice myself for this rubbing of yours.
Go below me and rub my body now
Water the garden of my vulva.
Press hard and squeeze into me."
The other said, "Cling to my neck
I am gasping, *dogana* dear,
Take a rest, for God's sake.
The water's begun to flow, stay and tell me,
Do say what is in your heart."
She said, "Ah, just like you,
I am in the same state as you."
When I heard their talk
I could not open my lips.
Shame prevented me from speaking
But I heard, and spoke to myself in my heart.
My husband had an affair with a person [*shakhs*]
But I was able to restrain my emotions.
I said to him, Come here
And listen to something interesting.
When that good-for-nothing came to me,
I showed him the women, and said,
My desires remained unfulfilled so I am sad.
This is not fit and proper, Rangeen.[25]

The speaker in this poem is abashed by the vigorous sex she witnesses, but also enticed by it, and wishes her desires could be fulfilled as completely as the two women's seem to be. Her husband is having an affair with a person whose gender is not revealed. In the last few lines, she reproaches him, arguing that it is not suitable for her alone to be left out of the lovemaking everyone else is enjoying.

Rangeen's second *Chaptinama* is narrated in the third person. It describes the anger of a man who witnesses two women's intimacy. At first he believes their pretence that they are just friends. But after he sees them having sex (it is possible that this is the same man who appears in the last lines of the first poem), he confronts them and insists that they have intercourse with him. He remarks that they are "loose women" anyway, and are in search of a man. He then has sex with them by turns all night until, worn out and bruised, they plead to be released. He then makes them promise never again to engage in *chapti*, otherwise he will continue intercourse with them as long as he has breath. They fold their hands, fall at his feet, and promise never again to "play *chapti*."

Critic C.M. Naim inaccurately states that the poet here depicts himself intervening and "curing" the two women by having sex with them.[26] In fact, the man in the poem is nowhere identified with the poet. The poem consistently uses the third-person pronoun *us* (he) for the male character. Whether he actually erases the women's desire for each other or not remains moot, because there is no evidence that they will keep the promise he extracts from them. The poet's only comment in his own persona is at the start of the poem (instead of at the end as is conventional), and casts doubt on the change the male character thinks he has effected in the women. The poet says:

> Sakht bedid hai ai Rangeen,
> Koi ab is ka kya ilaj karey?
> Mein usey chahun, chahey ghair ko yeh,
> Dosti isi tarah ki raj karey.
>
> (This is completely unseen, O Rangeen,
> How can anyone remedy it?
> I love someone who loves another
> May this type of friendship continue to reign.)[27]

Here, the poet suggests that sexual behavior, being invisible to others, cannot be controlled, and that sexual attraction is unpredictable. The tone is similar to that in the closing verses of the *Kamasutra* chapter on same-sex relations: "Practiced according to his fantasy and in secret, who can know where, when, how, and why he does it?" (II. 9. 45).

Gender Ambiguity

Some Urdu critics claim that the beloved's gender in the standard *ghazal* is always ambiguous, because the beloved represents God, who has no gender. However, other critics have conclusively shown that this is not the case. Details like the blouse, the breasts, the bodice, the eye holes in the veil, or the turban, cap, downy facial hair, or a direct use of words like *launda* (boy) often fix the gender of the beloved as female or male.[28] But verbs and other markers referring to him/her are always gendered male.[29]

The beloved in *rekhti* is always human and never divine, and is sometimes male and sometimes female. The *rekhti ghazal* thus conforms to a certain type of standard Urdu *ghazal*, except that its speaker is always female. Several *rekhti* poems keep the reader guessing the beloved's gender. The gender ambiguity of certain words in Urdu facilitates this. For example, *pari* (fairy) usually refers to a beautiful woman but in *ghazals* may also refer to a male youth:

> At night I heard a story of fairyland
> And in a dream saw the throne of Solomon.
> Can any fairy be as beautiful as you?
> I have never seen such a majestic person.
> That man is finished, is paralyzed and swoons
> Who feels the exciting heat of your thigh.
> This too is an ornament, like the rays of the sun
> When you wear a skirt of gold and silver brocade
> You told Nurse what should not have been told—
> May God never show me the face of one so naive as you. (KI 397: 2)

Here "person" and "fairy" leave gender ambiguous, but we are told that a man would be excited by this majestic man's or woman's thigh. The skirt (*lehnga*) reveals the beloved's gender as female, and the final couplet indicates that her liaison with the female speaker must be kept secret. Adding to the erotic ambience is the poem's allusion to the folk tale of Saif-ul Mulk and his fairy beloved, Badia ul-Zamal, the most beautiful girl in the world, whom he first saw in a vision, appearing out of Solomon's robe. The female speaker, having heard this story, declares her beloved more beautiful than any fairy.

Some *rekhti* poems never reveal the beloved's gender. Such is Insha's poem, where the speaker, in a fashion time-honored in Indic love poetry, laments the sorrows of love in the romantic monsoon season:

> Is the morning lying asleep in some garden somewhere?
> Why does the morning not arise in my presence?

If dark clouds had not gathered, O people,
I would not have lost my honor this morning.
The pearl drops that lie scattered on the greenery—
Did the morning bring a lapful of them to devote to me?
Sister, what is the use of the flower cups being washed—
Oh that the morning could wash the stains from my heart.
Oh gardener woman! Why does the morning not sow a seed
That will make everyone's hopes blossom and come to
harvest? (KI 408–409: 26)

Rekhti poems both please and tease the reader. In this Insha poem, ambiguity arises from the fact that "bulbul" and "mynah" are names of birds kept in cages as pets but are also terms of endearment used to address women:

I am devoted to you, my sweet one, don't scream
Don't wake up the sleeping people, dear, don't scream
....[editor's excision]
Don't wave your head and breathe so deeply, O bulbul,
I've already told you, *haan ri*, don't scream.
Why do you harass me, mynah, be quiet
Don't scream like a rustic girl, "Flown away, get away."
Don't meet Insha with screams and cries
Lady, I beg of you, I acknowledge defeat, don't scream. (KI 410: 29)

The speaker's injunctions not to scream may be addressed to a pet bird but may equally be addressed to a lover. This ambivalence persists to the end, with an erotic suggestiveness regarding screaming and deep breathing. These double entendres also appear in another poem where the addressee is a woman servant who is also probably the speaker's lover:

I tremble with fear at the thought of your plait
And wake startled at night, crying "A snake, a snake!"
Why did you come to the rooftop, O Dai, look,
You have woken all the sleeping people with your panting
. . . You say someone has found you out
Don't keep scaring me with your tales of being found out. . . .
(KI 404–405: 17)

The All-Women World of *Rekhti*

In the standard Urdu *ghazal*, the cast of characters is almost entirely male.[30] The poet's pen name appears in the last line of the *ghazal* and he or she conventionally identifies with the male lover-persona.

But in *rekhti* the cast is almost entirely female. The speaker is always female and she generally converses with another female. She has female friends and sometimes a female lover, female neighbors, and female rivals. She may have a husband and relatives such as co-wives and sisters-in-law. Children are rarely mentioned.

Where would one find such a predominantly female world in the late eighteenth and early nineteenth centuries in cities like Delhi, Hyderabad, and Lucknow? Not only in the women's quarters of a patriarchal family but also in a courtesan family.

Both in Western and in Indian literatures, there is a longstanding association between lesbianism and prostitution. While this is partly based on the negative stereotype of the over-sexed woman who, not content with "regular" sex, becomes a lesbian or a prostitute or both, it is also partly related to the material reality of prostitutes' lives. In many premodern societies, such as the ancient Greco-Roman world, and the medieval Indian Islamicate, courtesans were among the few women, apart from royalty, who had access to education and the arts, and could engage in free conversation with accomplished men. They also had relatively greater mobility and control over financial resources than most wives and daughters.

Both courtesans and lesbians live in ways that may allow them some degree of autonomy from male control; since both also may be seen as challenging the institution of heterosexual monogamy, they are often derided and punished in similar ways. Davesh Soneji's research on *devadasis* of Andhra Pradesh is suggestive in this regard, showing how social reform movements, organized by "respectable" women and men, force *devadasis* out of their positions in temples and outlaw their way of life. Nationalist social reform movements that arose in the nineteeth century under British rule tried to purify Indian life by outcasting non-normative sexual and relational arrangements, including *tawaifs*, *devadasis*, *hijras*, and same-sex relations.

Both in the West and in India, mainstream literature and cinema explore the connections between prostitutes and domestic women, which are denied and ignored by mainstream society. Victorian novels and poetry frequently expose the hypocrisy of men who father children by prostitutes, and convey venereal diseases to their wives. Modern Indian films too highlight the injustice of the divide between the world of prostitutes and that of the patriarchal family, showing that many prostitutes start off as wives and daughters in patriarchal families.

One commonalty between the world of prostitutes and the world of the patriarchal family that is rarely evident in mainstream literature in the late nineteenth and early twentieth centuries, is the pleasure of sexual relations, public in the former and private in the latter. Even in the domestic sphere, this pleasure may often be in excess of its normative procreative goal, but mainstream literature is generally too wary of this excess to acknowledge it non-punitively. Literature that nonjudgmentally explores this excess gets censored as obscene or labeled erotica.

Same-sex relations, being non-procreative, have historically almost always been viewed as excessive and their depiction relegated to the realm of erotica. In erotica, the barrier between the world of prostitutes and the world of domestic women often breaks down or is shown as permeable by the force of desire.

Erotic texts often depict female-female relations among prostitutes as episodic and a mere prelude to heterosexual sex. This is the case, for example, in John Cleland's novel *Fanny Hill* (1748–1749), where the protagonist Fanny is initiated into sex by fellow prostitute Phoebe's caresses, which prepare her for intercourse with a man. Similarly, the late Victorian underground magazine *The Pearl* always shows sex between women as a prelude or interlude to male-female sex.[31] But, as noted earlier with regard to Lucian's second-century Dialogue, some early texts do depict women's relations with each other independently of their relations with men. *Rekhti* does so too.

Courtesans and Same-Sex Unions

In Indian texts, the courtesan's household and the women's quarters of the normative household are two all-women spaces associated with female-female relationships. Although these spaces are conventionally seen as absolute antitheses of one another, female-female relations of various kinds constitute a sphere wherein the two spaces may also be seen as analogous. Thus, Yashodhara, in his *Jayamangala*, a twelfth-century commentary on the *Kamasutra*, while commenting on male homosexual relations, adds: "Women behave in the same way. Sometimes, in the secret of their inner rooms, with total trust in one another, they lick each other's vulva, just like whores."[32]

The idea that some prostitutes engage in lesbian relations is not just a literary invention. In her now-classic study of prostitutes in 1970s Lucknow, who are heirs to that city's tradition of accomplished courtesans under the Islamicate, Oldenburg found: "Almost every one of the women I interviewed during these many visits claimed that their closest emotional relationships were among themselves, and eight of them admitted when I pressed them, that their most satisfying physical involvements were with other women."[33] More recently, Shohini Ghosh's acclaimed documentary,

Tales of the Night Fairies, also alludes to close long-term relationships among the women sex workers of Calcutta whom she interviews.

In the introduction to his *rekhti* collection, Rangeen said that he picked up women's speech from the *khangis* of Delhi with whom he consorted in his youth. *Khangis* were married women who engaged in prostitution on the side, sometimes at home, sometimes in courtesans' houses. This category exposes the connections between the patriarchal household and the courtesan household. In several *rekhti* poems, the context seems to be that of a courtesan's house, where the speaker and her *dogana* adorn themselves not only for each others' but for a man's, sometimes the poet-persona's, pleasure:

> Your lap is full of flowers—well done, my *dogana*
> May your fields grow green—well done, my *dogana.*
> Hide me behind you and show me to that person today
> I am devoted to you, O my *dogana*, well done.
> ....[editor's excision]
> You wore green clothes to show them to Insha
> And became a green fairy, my *dogana*, well done. (KI 415: 42)

Rekhti shows that women in love experience emotions that are not restricted to any particular marital or social status. Medieval love poetry in general represents love and desire as equalizing forces that affect people similarly, across class and caste.

Urban Pleasures and Women's Language

> *Zaban ke khuld ki hai hoor aurat, Agar ho Lakhnau ke bostan se*
> *Zaban ka faisla hai auraton par, Ye baten marduey layen kahan se.*
>
> *(Woman is the houri in the paradise of speech,*
> *If [she and/or the speech be] from the garden of Lucknow.*
> *The choice of speech rests on women,*
> *From where can men produce such matters?)*
>
> *(Abid Mirza "Begam," IR 82)*

The women in *rekhti* are urban—the metaphorical "garden" omnipresent in Urdu poetry is not rural but urban. Historians have demonstrated that cultural representations of same-sex relationships generally occur in urban, not rural, contexts. City life facilitates mobility and increases the range of choices available to people.

Late medieval Indian cities under the Islamicate, like Delhi, Hyderabad, Agra and Lucknow, were hospitable to male networks of same-sex desire and love, celebrated in Persian and Urdu poetry.[34] Men who loved beautiful youths were identified by terms such as *husnaparast* (worshipers of beauty). Such men constituted an elaborate culture of lovers and beloveds, ideally absorbed in mystical devotion but as often engaged in more earthly relationships. Jur'at's female speakers Sukkho and Mukkho place their desires in the context of this wider range of urban pursuits, such as the pursuit of beauty: "Let's invite all the women in town who are given to *chapti* . . . All are absorbed in their own pursuits throughout the city/Waves arise in hearts that enjoy the river of beauty" (*SSLI* 222–223).

As opposed to the heavily Persianized Urdu of the mainstream *ghazal* at this time, *rekhti* uses the language of the streets and the women's quarters, an Urdu replete with words from other north Indian languages, such as Punjabi, Braj Bhasha, Avadhi, and Magadhi. *Rekhti* transforms this colloquial speech into a literary language, just as modern Hindi cinema makes the Hindi of the streets a cultural idiom.

Because Persian was the language of high culture, most Urdu poets composed in Persian. But if Persianized Urdu is their "father tongue," used in public spaces like the court, *rekhti* is closer to their "mother tongue"—the language spoken by their mothers and the female servants who raised them. It is also the language of their wives, and of the courtesans and male youths with whom they fell in love.

Rekhti poets, drawing attention to the elegance of their language, emphasize both its non-Persian ambience and its Indic urbanity. Thus, Jan Saheb writes: "Foreign aunt! You are a nightingale of Shiraz [in Persia]/ I am an Indian parrot and my tongue is eloquent/ . . . The wretched native hill crows cry 'caw, caw'/ I will hide my face if they can ever speak my language." (IR 59–60).

The most striking example of a *rekhti* poet claiming greater strength both for his own language as compared to standard Urdu, and for the female-female relationships he celebrates as compared to the male-male and male-female ones celebrated in standard Urdu *ghazals*, is this amazing couplet by Insha:

> Meri *dogana* aur main yun nahin hain jaisey rekhta
> Donon ki janein ek hain taney jo karti ho abas (406–07: 21)
>
> (My dogana and I are not scattered like Rekhta
> Our two lives are one; you taunt us for nothing.)

Through the pun on the word *rekhta* (which is a name for standard Urdu but literally means "scattered"), Insha points both to the pithiness of *rekhti*'s idiom and to its thematic innovation—the female lovers in *rekhti* poems are more often united than are the lover and beloved in standard Urdu (*rekhta*) poems. The idea of two lives becoming one through love suggests a marriage-like union.

Rekhti's Denouncers

After the British crushed the 1857 Indian rebellion, Muslim reformers began to purify Urdu by Persianizing its vocabulary. However, they also denounced Persian poetry for supposedly introducing male-male love into Urdu, and tried to Islamicize the language by purging it of erotic content, as well as of Hindu content they considered idolatrous. This was paralleled by Hindu nationalists developing Hindi as a heavily Sanskritized language, also purged of sexual content and of Islamic references.

Hindu and Muslim male reformers reacted to the British victory in 1858 by attacking cultural practices that signified the effeminacy the British attributed to Indian men.[35] Among these cultural phenomena were Hindi *reeti* and Urdu *rekhti* poetry, with their focus on women's worlds.

Rekhti poets often took female pen-names, such as "Begam," "Pari," and "Dogana." Some, like Jan Saheb, put on items of female apparel and mimicked female voices and gestures when they read their poems in poets' gatherings. Others like Insha, enacted different roles as they read different parts in the poems. Urdu critics in the late colonial period were uncomfortable with such practices typical of an indigenous urbanity laid waste by the British victory in 1858. Muhammad Husain Azad, a founding father of Urdu criticism, remarks that *rekhti*'s invention is "one cause of the effeminacy, lack of courage, and cowardice that grew up among the common people."[36]

In contrast, recent critics view *rekhti* poets as hypermasculine, assuming that they were heterosexual men who ventriloquized women, mocking and dehumanizing female sexuality in their poems.[37] Between those critics who see *rekhti* as obscene and those who term it patriarchal and misogynist, the genre has been downgraded in the canon, and many poems excised and lost.

I argue that literary representation cannot be reduced to authors' biographies and biological make up, which, in any case, are not entirely retrievable. Imaginative empathy between people who have different degrees of power, including men and women, is possible, and is often refracted through literary texts. A parallel to male authors of *rekhti* representing female voices is the widespread medieval phenomenon of mystic male poets identifying as females. An early example is seventh-century Tamil poet

Nammalvar, several of whose explicitly erotic poems are in the female voice.[38]

In my view, *rekhti*, by exploring female-female love, brings to the surface a dimension of women's humanity and sexuality that is present but not fully explicit in contemporaneous and earlier literary genres, like Hindi *reeti* poetry.

Rekhti and its Audience

Some recent commentators claim that the men who produced and enjoyed *rekhti* kept women in complete seclusion and excluded them from the Urdu literary tradition. Therefore women would have had no access to *rekhti*, and the female homosexuality it depicts reflects not women's experience but male voyeurism, mockery of women, and pornographic titillation.[39]

This argument wrongly assumes that all Muslim women in premodern Islamic urban India were secluded. As discussed earlier, many courtesans in Lucknow in the late eighteenth and early nineteenth centuries were accomplished singers, dancers, and poets, in whose salons educated men congregated to discuss literature and politics. In her examination of the civic tax ledgers of 1858–1877 in Lucknow, Oldenburg found that some courtesans were in the "highest tax bracket, with the largest individual incomes of anyone in the city."[40] Courtesans like Azizan Bai played a public role in the 1857 rebellion.

Although respectable Muslim women were not supposed to read or hear love poetry, many of them did read and even compose it. Kathryn Hansen quotes a late nineteenth-century biography that refers to women reading poetry: "having read Mir Hasan's *masnavi*, thousands of women became debauched . . . ."[41] Pritchett points out that several women, like the noblewoman Gunna Begam (died 1773), the princess of Bhopal, Shah Jahan Begam "Shirin" (1838–1901), and the dancing girl Mah Laqa Chanda (1766–1834), were well-known poets in their times. Historians like Azad excluded them (and also most Hindu poets) from the Urdu canon.[42] The exclusion, therefore, was not so much by other poets or readers in these women's own times as by later critics.

The seclusion of middle-class women was also not complete. Many housewives produced items for the market, such as fireworks, and had contact with marketing networks. There were also other non-secluded women, such as female innkeepers, vendors who carried goods to women in seclusion, and female soldiers and bodyguards maintained by wealthy persons such as dancing girl Mah Laqa Chanda, and even by the last king of Awadh, Wajid Ali Shah, who had his capital in Lucknow.

Urdu poetry was transmitted primarily orally, not through writing or print. It was recited aloud by poets in public gatherings as well as private intimate settings. Auditors learnt their favorite verses by heart and repeated them to others. This oral transmission could not be controlled or restricted the way manuscripts or books could be. An anecdote about a poet from Delhi conversing with a Lucknow courtesan named Bi Nuran is recounted by poet Insha and cited by Azad in his history of Urdu:

> Bi Nuran! . . . when you speak of reciting verses—there's no pleasure even in this any more, that you should make me recite. . . . And poor Mir Inshallah Khan, the son of Mir Masha'allah Khan, used to be a Parizad [fairy-faced one]—I too used to go to stare at him. Now lately he's gone and turned into a poet . . . let me tell you one more . . . His pen-name is Rangeen. He's composed a *qissah*. . . . in it he has used the language of whores. . . . All the people of Delhi and Lucknow, whether women or men, recite: "She went tripping away from there, lifting the hem of her skirt, /Causing her ankle-bracelets to tinkle together." . . . And because he consorted with prostitutes, there was a great deal of rakishness in Rangeen's character—so that he abandoned *Rekhtah* itself, and invented *rekhti!* So good men's daughters and daughters-in-law would read it and grow impassioned, and he could disgrace himself with them.[43]

Although based on hearsay like most such anecdotes, this nevertheless demonstrates that literary men were known to recite poetry (including *rekhti*) for courtesans. It also tells us that both women and men recited well-known erotic verses, and that *rekhti*, based on the language of prostitutes, was available to secluded women ("good men's daughters and daughters-in-law") who might "read it" and be seduced by it. Interesting too is the allegation that poet Insha was once a beautiful youth gazed at by other men. As this allegation indicates, *rekhti* poets lived in a society where it was not surprising for a man to have relations both with beautiful women and beautiful male youths.

From the late nineteenth century onward, British colonialism and Indian nationalism suppressed many gender and sexual variations, as well as literary genres that depicted these variations. However, the suppression was not complete, and the aspiration to same-sex union continued to express itself in fiction as well as modern cinema. Nineteenth-century *rekhti* poets were aware of the assault on their cultural world, and Insha comments specifically on the love that increasingly dared not tell its name:

> To take love's name is to get one's face scorched.
> How long can one burn in the heart's heat?
> Alas for this thought embedded in the heart
> How long to keep wringing these two hands?
> The *Dogana*'s voice is getting quenched today—
> Tell Insha, someone, that he should voice these sorrows now. (KI 432:78)

Chapter 9

Aspiring to Union: Twentieth-Century Cinema

Tum jo huey mere humsafar,
Raste badal gaye,
Lakhon diye mere pyar ki
Rahon mein jal gaye . . .
(Now that you have become my companion on the journey,
All the roads have changed,
Millions of lamps
Light up the paths of my love . . .)

—*Song from* 12'o'clock *(1958)*

When I was a feminist activist in Delhi in the 1980s, I looked for and found gay friends who met socially in each others' houses. One thing that surprised me about this little community was its immersion in Hindi film songs. Every party, small or big, climaxed in individual or group rendition of film songs, mostly romantic numbers from pre-1970s cinema. Film songs are ubiquitous in India. It is nearly impossible to be in a public space or most private spaces without a song drifting in from someone's radio, TV or player. But since popular cinema, like Indian society in general, is obsessed with heterosexual marriage, and almost every film culminates in weddings or deaths that forestall weddings, I could not understand why gay people were so invested in what seemed to me a heterosexist cultural product.

Over the next few years, though, Hindi film songs entwined themselves with my relationships—there seemed to be a song for every mood and moment in the trajectory of love. And I realized, as I listened to, hummed, or sang these songs, that many of them are not gendered. They float free of the films from which they come, and may now be sung by anyone, regardless of gender or sexuality—in our circle, a gay man was the best singer of songs originally written for women. Many people do not remember the films in which these songs were originally performed. Sung by famous singers and re-sung by many others, the songs have a much longer life than the films they are part of.

In this chapter I focus on Hindi cinematic traditions of portraying same-sex relationships within an overall romantic and erotic economy that privileges love and commitment, regardless of gender. Bombay cinema, although in Hindi, represents perhaps the closest thing to an all-India cultural language that exists today. Its speech ranges from chaste Urdu to a complex mix of Hindi, English, and other Indian languages and dialects that mirrors the lingo of the urban streets. It appeals to pan-Indian audiences, cutting across religious, regional, gender, class, and linguistic lines, and is even popular in many West Asian and Southeast Asian countries. Like cricket but more so, it produces cultural heroes (actors, singers, song writers) who are Muslims but retain vast fan followings in all communities, including the Hindu community, even in times of intense Hindu–Muslim conflict. It is a primary modern carrier of the Indic traditions I explore in this book.

Hindi cinema has shaped many modern Indian practices, including wedding practices. The kind of celebrations it depicts, ranging from bridal apparel to collective dancing and singing, strongly influences people's choices when they plan their weddings. Hindi film music is the type of music most often played at wedding celebrations. Weddings are central to Hindi films—my question is: how do same-sex unions figure into this dynamic?

Gender and Romantic Songs

Many songs, especially older ones, are in the first and second person ("I love you") and thus carefully avoid gendering either singer or addressee. Take, for example, a famous, deeply romantic song from the 1950s: "The earth is silent, the moon and stars are silent, my heartbeats call out to you . . ."[1] This way of avoiding gender is a time-honored strategy adopted by poets in many cultures to mask same-sex love and also to make their poems accessible to all, regardless of gender. Lord Byron and W.H. Auden, among many other male poets in English, wrote poems to young men, phrasing them in the "I-you" mode. Many Valentine's Day cards adopt the same strategy today and thus become more marketable—a woman or a man can buy them for a woman or a man.

Some of the most romantic songs in the Western corpus are also written in this ungendered mode. "Drink to me only with thine eyes," for example, does not reveal lover or beloved's gender. Although Ben Jonson (1572–1637), its author in English, titled it "To Celia," the song is, as any educated person in Jonson's England would have known, a pastiche of translations from ancient love letters and poems written by men for male youths.[2]

Hindi songs also use the optative voice and the plural to avoid gendering speaker and addressee. Even those songs that are gendered or sung as duets by a male and a female voice, transpose easily to non-heterosexual contexts, because of their overwhelming emphasis on love as an emotion experienced similarly by all rather than on normative gender roles.

This openness appeals to gay people. Many Indian gay men say that they identify with female film stars, like Meena Kumari, Madhubala, Helen, and Rekha, who are known for their bold screen personae as well as their off-screen romantic lives. Some Indian lesbians have told me that they identify with Don Juan type male stars like Dev Anand.

Love—the All-Inclusive Dream

Shohini Ghosh has suggested that one reason Indian cinema is attractive to queer subcultures is "its privileging of romantic love as the most important of all emotions."[3] I would modify this to argue that Indian cinema projects love, not just romantic love, as the most important emotion. It could be love for a friend, a mother, a child, or a lover. Love of all kinds constitutes a space that aspires to include everyone, even villains and outcasts.[4]

This phenomenon is common to Hollywood and Bombay cinema. In an interview with gay magazine *The Advocate*, Barbra Streisand, who is herself heterosexual but has a devoted gay male fan following, remembers that popular movies helped her, a lonely teenager who thought she was unattractive, to be hopeful about finding love: "Movie music is romantic. The music in certain films is what makes you cry."[5]

Popular cinema, both Hollywood and Indian, draws on age-old conventions to figure Eros or Kama as an irresistible force that can strike unexpectedly at any time but cannot be compelled into existence, simultaneously suggesting that this apparent chaos is ordered, because destiny arranges the right match at the right time. For instance, *Sleepless in Seattle* (1993) develops the idea of love as a dream—many scenes occur at night, and chance encounters acquire an almost mystical character when the hero and heroine are inexplicably drawn to one another. The hero's child acts as the instrument of destiny, and circumstances collaborate to unite the couple despite all obstacles. In *When Harry Met Sally* (1989), the dreamlike quality of love is replaced by an apparently more realistic development of friendship between hero and heroine, but the spectator knows throughout that the two are "meant" for one another.

Hindi cinema's love plot may have a conformist conclusion, but throughout the movie it usually fosters defiance of convention and protest against injustice. Unconventional love matches—cross-caste, interreligious, interregional marriage, and widow and divorcee remarriage—have all been sympathetically portrayed, while cross-class marriage is practically mandatory. Lovers are routinely shown resisting forces that try to separate them, from family to police to government. They prefer torture and death to separation. In *Mughal-e-Azam*, 1960, Prince Salim defies his father, Akbar, to pursue a romance with a dancing girl. She sings a famous song in public defiance of the king. Its refrain is: "When in love why fear anyone? It's love, not a theft." A friend tells me that he has heard lines from this song frequently quoted by gay people: "When there is no *parda* (veil) from God, why keep *parda* from humans?" This message is bound to appeal to those forced to keep their love secret.

Chosen Family in Popular Cinema

Finally, popular cinema is attractive to gay people because it celebrates nonbiological family ties. While the right wing in both the United States and India extols family values as if all families are exactly the same, literature and cinema more realistically portray a range of family arrangements. Protagonists are often raised by single or paired uncles, aunts, siblings, friends, and other adoptive parents, who prove vastly superior to biological parents.

This type of representation has a long history in the West. Take, for example, the shepherds in the ancient Greek play *Oedipus Rex*, and in Shakespeare's *The Winter's Tale*, who adopt babies abandoned by biological parents. Most heroes and heroines in Victorian novels are raised in unusual families. Fanny Price in Jane Austen's *Mansfield Park* (1814) realizes that the aunt and uncle who adopted her are better parents than her birth parents; the servant Nelly in Emily Bronte's *Wuthering Heights* (1847) raises more than one generation of children mistreated by biological parents; and Dickens' David Copperfield, whose mother cannot protect him from abuse by his stepfather, finds a true family with his spinster aunt Betsy Trotwood and her odd partner Mr. Dick, as does the orphaned Oliver Twist with the bachelor Mr. Brownlow. This pattern is perhaps most famously dramatized in George Eliot's *Silas Marner* (1861), when Eppie chooses her adoptive father, a poor weaver, over the wealthy biological father who abandoned her as a baby.

Hollywood continues the tradition of portraying odd parents sympathetically—aunts, uncles, adoptive parents are often shown as far better than biological parents. A recent example is *Secondhand Lions* (2003), where a child's two great-uncles, eccentric bachelors, become the nurturing family his mother could never provide.

Hollywood also celebrates friends as constituting chosen kinship. An important genre of feel-good American movies, as of TV sitcoms, shows the biological family as dysfunctional or disrupted, and a chosen family developing to supplement it. *Under the Tuscan Sun* (2003) is characteristic—an American woman abandoned by her husband for a younger woman, sets up house in Italy, where she befriends a neighboring family, presides over the romance of a young immigrant with a local girl, and offers shelter to her pregnant lesbian friend.

Despite the stereotype of the traditional Indian family, popular Indian movies, many of them based on novels and short stories by Indian writers, follow a similar trajectory. Many films sold as "family movies" depict families headed by widowed or divorced single parents or consisting of orphaned siblings. A famous example is *Mother India* (1957), in which a heroic peasant woman raises her sons on her own, when her crippled husband abandons the family. Almost all of superstar Amitabh Bachchan's megahits depict him as the product of a disturbed family or an orphan and protector of orphans. In *Lawaris* (Orphan), 1981, he is an illegitimate child abandoned by his father and raised by a drunkard after his mother dies; in *Trishul* (Trident), 1978, he is an illegitimate child raised by his mother; in *Zanjeer* (Fetters), 1973, he witnesses his parents' murder; in *Coolie*, he is a Muslim, orphaned when the villain kills his father; in *Namak Halal* (Loyal One), 1982, he is given up by his mother and raised by his grandfather; in *Don*, 1978, he adopts two children separated from their father. In *Hum Aapke Hain Kaun* (What Am I To You?), 1994, a paean to the joint family, a bachelor uncle raises the orphaned heroes. Friends too figure prominently, as an almost indispensable part of a happy life. As discussed in chapters 1 and 6, this pattern derives from premodern narrative.

Film Songs and Women's Eroticism: Picking Up Where *Rekhti* Left Off

Ghosh and others argue that pre-1990s film heroines were divided into "the westernized," sexualized vamp and the "chaste" heroine who subordinates sexual desire to love and duty, only occasionally masquerading as a sexualized woman.[6] This formulation splits desire from love and also omits the erotics of work and dress. Hindi film heroines engage in a wide range of adventures and occupations even in the1940s and 1950s, Nadia being the epitome of this adventurousness. It was not just the vamp, but even the heroine who wore trousers, cut her hair short, lived and worked on her own, and had forbidden romantic relationships.

More important, like *rekhti*, Hindi film songs celebrate women's erotic being in women's own voices. While some old film songs depict women lamenting rejection, like the male lover in the *ghazal*, many others show women awakening autonomously to their sexual feelings, and then asserting them in relationships. For example, in a lovely song from the 1950s still hugely popular today, a girl expresses her budding sexuality in the context of the romantic monsoon season; many lines dwell on her autoerotic emotions: "Today I grew shy, looking at my own reflection . . . /A lock of my hair tumbled with joy, having kissed my lips . . . /As I set out, my heart said,/ Sway even more, /My being, full of love, romped and laughed, /Turned around and danced . . ."[7]

In several songs, women court men they desire. In one from the 1960s, a woman sings, "The night is alone, the lamps are out, /Come close to me and whisper in my ears whatever you want . . . You may not love me, but I love you and I give you love's permission, /Why be afraid? /Say whatever you want."[8] In another, the heroine teases her lover for his simplicity, "The moon blossoms, the stars laugh, /This night is strangely intoxicating, /Those with understanding have understood, /One who does not understand is a simpleton indeed."[9]

While the plots of most films are heavily didactic and censorious of erotic pleasure, the half-dozen songs that punctuate every film tend to celebrate the erotic life, including illicit eroticism. Songs articulate the inner life, and stand in for erotic/emotional life. Judged by standards of "realism," characters burst into song at inopportune moments, when people would more likely be silently overwhelmed by emotion. But in Indian cinematic convention, these are the most opportune moments for song, because the audience understands songs as devices for articulating emotions. Often, these are erotic emotions. Frequently, a song and dance sequence is presented as the dream of one character, which may or may not be translated into action. Women characters may not have a happy sex life but even the most oppressed are often allowed a glorious fantasy through a song.

Film songs thus continue Indic traditions of celebrating the erotic life, even while the plot lines record the anxieties attendant on that life, especially the anxieties endemic to modernity. Like *rekhti* and the Indic erotic traditions in which it is embedded, film songs often nonjudgmentally celebrate hedonism: "O my life, dance tonight, kiss the sky, Who knows whether or not tomorrow will come?"[10]

As in *rekhti*, the lyrics for film songs are almost all written by men, and in the early decades, often by major Urdu poets, like Josh Malihabadi, Shailendra, Shakeel Badayuni, Jan Nisar Akhtar, Sahir Ludhianvi, and Majrooh Sultanpuri. This tradition continues with Kaifi Azmi, Gulzar, Javed Akhtar, Neeraj, Sheharyar, and others. Just as *rekhti* was written in the female voice, men write film songs to be sung by women singers and performed by actresses. It has often been noticed that Hindi film songs draw on traditions of love poetry, especially the *ghazal*, both in form and content. I would add that the erotics of genres like *rekhti* and *reeti* poetry, pushed underground, resurfaces in film songs written for female voices. For all these reasons, the aesthetic value of songs often far surpasses that of the films in which they appear.

Love in a High Camp Tone

Hindi film songs can be serious and lachrymose like the mainstream *ghazal* but they can also be tongue-in-cheek and playful like *rekhti*, parodying the melodrama of romance even while participating in it. This makes the songs hospitable to gay improvisation and adaptation.

An old example is a still popular song, which plays on the trope of the lover losing his/her heart, except that the organ in question here is the liver, the seat of passion in medieval erotics, both Indian and European:

> I don't know where my liver has gone
> It was right here, where has it gone?
> It has died, overcome by somebody's airs,
> It took fright at someone's large eyes.[11]

A man and woman engaged in an office romance scramble under desks and chairs, searching for the lost liver, as they sing this song, concluding that they must report the loss to the police.

The Lesbian Moment: The Successors of *Rekhti*

Cinema follows premodern literary convention, focusing on male-male bonding, and neglecting female-female bonding. There are some brief scenes suggestive of lesbian eroticism, as in *Rajnigandha* (1974), and *Razia*

Sultana (1983). In the Marathi film *Umbartha*, remade as the Hindi *Subah* (Morning, 1983), a women's prison warden speaks up for two lesbian prisoners when everyone else is shocked by their relationship. This is perhaps the only instance of explicit lesbianism in Bombay cinema, until *Girlfriend* (2004), which follows an outdated American 1950s formula, depicting a lesbian as a murderous lunatic, and her girlfriend as a heterosexual in denial.

Filhaal (In the Meantime, 2002), directed by Meghna Gulzar, is premised on female bonding, because the married, infertile heroine's best friend, a single woman, agrees to act as surrogate mother for the married couple's child. But the bond lacks conviction, and rapidly degenerates into jealousy and conflict, with the male-female relationships proving much more supportive.

Lesbian eroticism surfaces in song and dance numbers in all-women settings, where girls cross-dress and/or flirt with each other.[12] The playful erotic economy in this genre of songs, which are found from the earliest films to the latest, is reminiscent of that in *rekhti*—take, for example, the still popular song, "*Reshmi salwar kurta jali ka*" from *Naya Daur* (1957). Two dancing girls in a theatrical troupe visiting a village perform this song. The playback singers are Shamshad Begum and Asha Bhonsle, the first of whom came from a courtesan lineage. The ambience is that of a disappearing sexual economy in which courtesans and dancing girls play central roles. One dancer wears men's clothing but it does not function to make her manly. Her breasts are prominent, and her girlishness evident. In the song, she uses the masculine first person pronoun, but the playback singer's voice is clearly feminine.

The song is highly flirtatious, as one girl makes advances to the other, who coyly resists and then responds. As in *rekhti*, the singer details the effect on her of the other girl's dress and beauty: "A *salwar* of silk and a *kurta* of net, /The delicate one's beauty is too much to bear/Whenever I look at you,/Fireworks go off in my breast." When the beloved playfully threatens to report her to the police, she professes willingness to be handcuffed.

Interesting too is the largely male audience mesmerized by this display, which is framed by the stage and thus at one remove from them, as the *rekhti* text is from its auditors. Village women constitute part of the audience but the camera does not focus on them, except when the heroine stands at the theater doorway, almost at a second remove from the text, like the female reader of *rekhti*.

Love and Friendship Rewritten for Modernity

Indian cinema appears in the era when the premodern ideal of same-sex friendship is giving way to the modern ideal of heterosexual romantic coupledom, and it reflects this shift. As discussed in chapter 1, movies such as *Yaarana* (Friendship, 1981) show a girlfriend supplanting the primary intimacy between two men.

But the shift is not complete. In contrast to modern Indian fiction, which relegates male homosexuality to the underworld and focuses on female homosexuality, cinema continues an older tradition of celebrating male-male romantic friendship. With a few exceptions, such as *Holi* (1984), which showed male students abusing a gay classmate and driving him to suicide, Hindi cinema rarely shows the dark underside of violence against gay men.

Male-male friendship in films up to the 1990s continues to carry as much emotional intensity as male-female romance, and often more. Male friends express themselves in songs as romantic as those sung by male-female couples. In almost all such films, however, the two friends end up married to women; alternatively, one marries a woman and the other dies, as in *Anand* (Joy, 1970).[13]

Although union with a male friend rarely appears as a long-term option, this union nevertheless remains an ideal. A good example is *Sholay* (Embers, 1975), a film often credited with inaugurating a new era in Hindi cinema. Here protagonist Jai's male friend Veeru and girlfriend Basanti both declare their willingness to die for him, but it is Veeru who actually does so. The villain terms male-female love "*yaarana*," the word also used for the male-male bond. Jai and Veeru sing a duet: "We will never break this friendship,/ I will die but never leave you. /Your victory is my victory, your defeat my defeat, /O my *yar*, /Your sorrows are mine, /My life is yours, /Such is our love . . . / We look like two to others but we are not two . . . / We eat and drink together, /We live and die together . . . " This is indistinguishable in sentiment from the heroine's love song to Veeru: "Love never dies, never fears death, /Even if we are destroyed and die, /Our story will live on." As he dies in Veeru's arms, Jai tells Veeru never to forget the story of their friendship and to tell it to his children, and says he does not regret dying because "I lived with my friend and died in my friend's presence."

There are, however, a few films in which it is possible for two men to live together instead of dying for each other. One is *Dosti* (Friendship, 1964). Thirty years later, the formula was updated in *Tamanna* (Longing, 1997).

Same-Sex Unions

In these two films, male-male relationships have the hallmarks of committed unions—they are primary, exclusive, long-lasting, intense, and they foster moral virtue. In *Dosti*, the protagonists' virtue is signaled by their names, which are the names of Hindu Gods—Ramu and Mohan. They are also marked as victim-martyrs by their disabilities—one is lame and the other blind. Their relationship fulfils all the criteria discussed in chapter 5 as indicative of true love/friendship: likeness, sharing, rituals and vows, sacrifice, and suffering for one another. The relationship develops in the context of a chosen family and ends with a public acknowledgment of their union.

Intensity between Ramu and Mohan is figured through conventions of poetic speech, song, and narrative (see photo 9.1). Soon after they meet, Ramu admires Mohan's eyes: "Such beautiful eyes . . ." and a little later, repeats, "Truly, how beautiful your eyes are." The relationship rapidly progresses to a vow of sharing when Mohan refuses to eat without Ramu: "Look, once you have taken my hand don't consider me a stranger." The ritual of holding hands is followed by eating together, wiping each other's tears and making music together (Ramu plays the mouth organ while Mohan sings).

The intensity is heightened by shared ordeals. Bullies abuse, mock and rob them, and the rich brother of a little girl they befriend despises them. They assemble a chosen family—the little girl, a motherly neighbor, and Ramu's schoolteacher. The last, however, forces a separation between the boys. This turning point deepens the relationship's intensity.

Photo 9.1 Mohan (left) and Ramu, in *Dosti*, 1967. Photo: *National Film Archive of India.*

Dosti became a hit largely on the strength of its songs. Composed by poet Majrooh Sultanpuri and sung by Lata and Mohammad Rafi, the songs have canonical status today. Mohan's second song presents their relationship as consolation for sorrow: "We may be far from our destination/ But our love is enough for us, /Even if thorns prick our feet/This support is enough for us." Putting his arm around the weeping Ramu, he continues: "At least your companion on the journey (*humrah*) is someone of your own."

Mohan's song, "I will love you morning and night, but will never call your name again" is a good example of how the ungendered love song functions out of context. Played on radio and television, it sounds like the lament of the conventional romantic lover, because the second person ("I" and "you") is used throughout, and the words for "beloved" and "friend" such as "*yar*" and "*mitva*," although gendered male, are conventionally used for a beloved of either sex. The word used for the boys' love both in this song and elsewhere in the film is "*chah*," literally "desire." The song continues on the high road of romantic love, identifying the beloved as the lover's be-all and end-all: "You are my pain, my rest, my eyes . . ."

Mohan proves his love by suffering for his beloved's welfare. When the schoolteacher dies, leaving Ramu unable to pay his examination fee, Mohan rises from his sickbed and goes out to earn by singing. He walks barefoot through the rain, singing, "Whatever step I take/ Is on your path, /Because wherever you are/ I am watching over you." The song goes on to restate the basic definition of love in Hindi cinema and in many older traditions, both Indian and Western: "The bond of pain is true/ What separation then? / Only they can be separated/ Whose love is false."

The film ends with a grand reunion, Ramu falling into the ailing Mohan's arms and asking forgiveness. Ramu declares: "No one can separate us now, Mohan," to which the elderly neighbor responds, touching both their faces: "May your enemies get separated. May God keep the pair of you united forever."

This final blessing by an older mother figure is charged with cultural resonances, and is impossible to translate literally: "*Bhagvan karey tumhari jodi isi tarah bani rahey*" (May God keep the pair of you united thus forever). "*Jodi*" is a couple or pair. This blessing is one that elders traditionally give married couples and is thus suitable for the end of a Hindi film, except that the pair here is male.

The Adult Male Pair

Ramu and Mohan are adolescent boys, but the protagonists of *Tamanna* (1997) are middle-aged bachelors, both Muslim. The greater self-consciousness of this film is signaled in its figuring the protagonist, Tikkoo, as a *hijra*. We are told that he was born a hermaphrodite. He and others see this as a disability, but it also signals his deviant gender/sexual identity.

The film is based on the true story of a *hijra* named Tikkoo, who worked in the film industry, and adopted a girl baby. But the real-life Tikkoo never attempted to "pass" as a man. The film protagonist Tikkoo's attempts to "pass" as a man create emotional drama when his adopted daughter discovers his "true" identity as a *hijra*. The film thus uses the tropes of "closeting" and "outing," which are relevant to homosexual people in India today who often lead double lives, but not as much to *hijras* who usually publicly display their difference.

In the film, other *hijras* live, as most do in modern India, in a group, dressed as women, using female names and personae, but Tikkoo has an aversion to them and asks them to stay away from him. They, however, remain sympathetic to him, and help him and Tamanna at crucial junctures. He always dresses as a man, all his friends are men, and he has a long-term unmarried male companion, Salim, with whom he raises two children. His body language and camp mannerisms are more suggestive of the self-presentation of many urban gay men in India today than that of *hijras*. By labeling him a *hijra* the film avoids direct discussion of homosexuality (this distinction is complicated, however, by the fact that many homosexually inclined men of lower-income groups in India today join or live on the fringes of the *hijra* community). The film's representation of him allows slippage between the *hijra* persona and that of a gay man.

Tamanna focuses on Tikkoo's love for his adopted daughter Tamanna, with the love between Tikkoo and Salim as an understated but ever-present backdrop. The implicit masculine-feminine coding here, the bearded, silent and gruff Salim supporting the long-haired, dramatic, often hysterical Tikkoo, is clearer than in *Dosti*, where the football-playing, school-going Ramu is fostered by the sensitive, fair, and delicate Mohan. Tikkoo's emotionality, which initially marks him as "not-man," also ultimately marks him as the true "man"—the true human being.

Same-Sex Parenting

Parenting of adopted children, in *Tamanna*, is shown as far more selfless than parenting of biological children. The film's theme song, sung by Sufis, points out that to cherish a child is to serve both the community and God: "The mosque is very far from home, let us make a crying child laugh."

In *Tamanna*, made in the 1990s when the gay movement was becoming increasingly visible in India, parenting also works as a safety device, deflecting possible anxiety about the homosexual implications of the relationship. Unlike Ramu and Mohan, Salim and Tikkoo are not shown sharing a house or a bed (though they are always together and Tikkoo says he often sleeps at Salim's house), nor do they sing songs about their love for each other. Their bonding through co-parenting establishes their love as normative and them as ideal men.

Co-parenting also has the effect of cementing the men's coupledom, which repeatedly becomes a visual focus on screen. At the film's climactic moment when Tamanna sees Tikkoo dressed as a *hijra* (he has forced himself to dance with other hijras to earn money for her education) and rejects him, saying he cannot be her father, an enraged Salim slaps her and cries, "He stayed awake nights clasping you to his chest so that you could sleep, he walked barefoot to feed you." He then leads Tikkoo off by the hand, muttering in standard parental fashion: "These children of today . . ."

Later, when both Tikkoo and Tamanna weep in despair, Tamanna's boyfriend consoles her, clasping her with his arm, while Salim similarly clasps Tikkoo a short distance away. At the film's conclusion, when Tamanna chooses to stay with her adoptive family rather than go to her biological family, the same formation recurs, with each of the two couples (Tamanna and boyfriend, Tikkoo and Salim) embracing.

The Tikkoo-Salim couple is exclusive and primary since neither has any heterosexual involvement or any other close male friend. Their relationship is shown to last from young adulthood to old age. While the youths in *Dosti* are bonded to the community as son and brother figures, the men in *Tamanna* are bonded as parent figures. In both films male-male love is demonstrated to be socially useful and outreaching rather than inward-turning.

Virtue, Civilization, and Same-Sex Union

In several Hindi films, devoted same-sex union signifies the virtues of civilization and humanity; in several other films, filial, wifely or parental devotion has the same effect. As in medieval European romances, same-sex relationships appear even more selfless than familial ones, because the partners do not have any material expectations of one another.

Chaudahvin ka Chand (Full Moon), 1960, is a classic, best known for its title song, in which the hero praises the heroine's moonlike beauty. However, the plot is driven by two male friends' love for the same woman, and their greater attachment to each other, which impels each to sacrifice his love for the other.

The film is set in Lucknow, a city figured nostalgically as typifying Indian culture. The film's first song, a panegyric to Lucknow, sets the tone. Its culminating stanza, marked as special by a dramatic slowing of tempo, praises friendship as the symbol of Lucknow's, and implicitly India's, greatness. The song is ungendered; it conflates friendship and romantic love, using conventional words for love like *mohabbat* to characterize friendship: "This land of Lucknow, /Here friends are ever faithful, /Full of love, /Once they become another's, /They remain theirs for life. /Maintaining their own dignity, /They expand the heart's grandeur . . . / If asked, they will even give up their lives, /If they have faith in friendship,/This land of Lucknow . . ."

Films like *Dosti* and *Tamanna* dramatize the sentiments of this song. It is important that, as in *rekhti*, all three films are set in cities—urban and urbane culture is central to the values they celebrate. Civilization, as the word indicates, is connected to city life. In the city, the marginal and abnormal—the cripple, the blind, the *hijra*—are the true heroes who make sacrifices to preserve cultural values.

In *Tamanna*, Tikkoo initially appears "abnormal" or subnormal but rapidly develops into a supernormal person who is almost divine. When Tikkoo continues to act with loving forbearance toward his wealthy half-brother who insults and despises him, Salim starts crying. This is the only time he cries, whereas Tikkoo, like the typical Hindi film heroine, cries frequently. Salim bursts out, "Tikkoo, don't place this great debt of your goodness on me. I cannot bear this burden, spare me. Make your heart a little smaller for a while; become human for once. I don't want to worship you, I must worship God."

In many religious narratives (such as the Biblical conversation in Genesis between God and Abraham, where Abraham asks God to spare Sodom), God is willing to spare the human race despite its wickedness, for the sake of a few virtuous people or even one virtuous person. In Hindu legend, the divine manifests itself in virtuous beings, human and non-human. In *Dosti* and *Tamanna*, this virtuous person is initially marked as "less-than-man" but turns out to be a truer man or human than others. As Salim says of Tikkoo, "If he is a *hijra* [here used to signify non-man] then shame on us men."

Tikkoo, despite his non-normative gender status, comes to stand for normative Indianness as well as normative humanity. It is highly significant that both he and Salim are Muslims and thus doubly marginalized; the baby girl they adopt was abandoned by the stereotypical oppressive patriarch of a Hindu family. A *hijra* briefly plays a similar role in Bombay (1995), when, like the hero, he stands between a mob and its intended victims, during a Hindu–Muslim riot.

Gay Subtexts: An End to Male–Male Romance?

From the late 1990s onward, the tradition of celebrating male-male non-sexual romance is dealt near-fatal blows in movies that include gay subtexts. It has become almost mandatory to have a minor gay character, usually an exaggeratedly effeminate male, who either is or is suspected of being gay (*Muskaan*, 2003; *Out of Control*, 2003).[14] Gay jokes abound: for example, in *Kal Ho Na Ho* (2003) an Indian maidservant in New York misreads two friends as lovers. The director, who is widely rumored to be gay, and both male actors played up this joke when hosting a televised awards ceremony.

These subtexts, however, tend to erode the passionate romantic bonding characteristic of Hindi cinema. A good example is *Mein Khiladi Tu Anadi* (I am a Player, You are Unskilled, 1994). In a self-reflexive move, the film tracks the efforts of Deepak, a romantic, old-style movie star to remake himself as a macho, Bachchan-type action hero. He trains himself by latching on to Karan, a hard-as-nails policeman who despises him as a sissy. Deepak tells Karan, "I am mad about you," but his attachment is silly and self-serving, and becomes emotionally convincing only when he falls in love with Karan's sister. The traditional motif of friends sacrificing for each other is parodied when Karan pretends to sacrifice his career for Deepak, his real motive being to get Deepak out of his and his sister's life.

Despite or because of the film's many homosexual double-entendres, the male-male bond never acquires intensity, and the song celebrating it emphasizes that heterosexuality is all they have in common. Its refrain is, "We both are very different, /I am a player, you are unskilled," but the verses show them chasing girls with crude gusto: "When I see a girl my heart beats, /I whistle and clap my hands . . ."

More positive portrayals of gay people, along the lines of *Fire*, occur in Mahesh Dattani's film *Mango Souffle* (2003), based on his award-winning play, and *Rules: Pyar ka Superhit Formula* (2003). In the latter, one of the six romantic couples is male. Here, finally, is a portrayal of how cross-sex marriage interferes with same-sex unions. The parents of one of the men try to arrange his marriage. While they are introducing the prospective groom to his bride, the groom's real spouse roams around, distraught. He stumbles into the drawing room where the families are, during a power cut, and the two men kiss as the power returns and the lights go on. The on-screen kiss, rare and, until recently, taboo in Hindi films, makes visible an aspect of male-male romantic friendship that has so far been in the dark, and disrupts the normative cross-sex marriage plot.

Since Hindi cinema, especially in its music, is hospitable to love bonds that break taboos, and since many major players in the world of film and theater are gay, hopefully, movies will in future more openly acknowledge same-sex marriage-like unions that have so far been celebrated as friendships. As a song from *Hum Aapke Hain Kaun* puts it, *Sab rasmon se badi hai jag mein, /Dil se dil ke sagai* ("The greatest of all rites in the world, /Is the engagement of heart to heart").

Chapter 10

Conclusion

Relationship is the mirror in which we see ourselves as we really are.

—*Jiddu Krishnamurti*

The only thing that can be premised with certainty about human nature is that it changes.

—*Oscar Wilde*

Marriage: the legal or religious union of two people.

—*Canadian Oxford Dictionary*

On April 1, 2001, the first legally recognized same-sex marriage in the world took place in the Netherlands. In the act of marrying, same-sex couples declare the moral and social equivalence of their relationships to cross-sex relationships. They change not only their lives but also their societies.

I argue throughout this book that couples who go through the rites of marriage, with the benefit of some social recognition, are in fact married, whether or not the law recognizes it. Marriage, as its history shows, is not the exclusive domain of the state. India's marriage law is relatively more pluralistic than marriage law in most Western democracies. Because the Indian government recognizes any marriage performed according to ceremonies customary in one partner's community, I have argued that same-sex weddings, performed with family approval and in some cases by priests, and acknowledged by the partners' communities, are valid marriages, and function as such in society.

Gay Samaj

However, in the few cases where Indian couples have sought government recognition for their marriages, local authorities have refused that recognition, and no couple so far has taken the issue to court. The recent remarks of some authorities on traditional Hindu law support my contention that these marriages are legally valid.

At the Kumbha Mela (major annual pilgrimage) in 2004, *Hinduism Today* reporter Rajiv Malik asked several Hindu Swamis their opinion of same-sex marriage. Pandit Shailendra Shri Sheshnarayan Ji Vaidyaka of *avahan akhara* (an *akhara* is a monastic order) pointed out: "Whatever is done in a hidden manner becomes a wrong act and is treated as a sin. But whatever is done openly does invite criticism for some time but ultimately gains acceptance. Why not give them the liberty to live in their own way, if they are going to do it anyway? After all, we have *kinnars*, eunuchs, who have been accepted by the society. Similarly these people can also be accepted. Like we have a *kinnarsamaj*, eunuch society, we can have a gay *samaj*."[1]

This statement supports my contention that just as relatively new communities (such as the Arya Samaj and the Self-Respect Movement) acquire legal recognition for the wedding ceremonies they conduct, the gay community too can, if it performs several weddings over time. An example is a gay community in Baroda, Gujarat. Sylvester Merchant, 22, an officer of Lakshya Trust, a Baroda gay organization, says, "In four years, we have facilitated at least 15 gay marriages and soon hope to introduce a gay couple club to extend emotional support. . . . The marriages will promote single-partner sex, which will help the HIV prevention campaign."[2] One male couple in their early twenties got married by both Hindu and Muslim rites. Another couple, Sandeep, 21, and Raj, 28, had a Hindu wedding in September 2004, attended by 200 guests.

Baroda is the city where two female couples registered their unions in 1987 (see chapter 4). It is also the hometown of Bhupen Khakhar (1934–2003), one of India's greatest painters, who came out as gay in the 1980s, and many of whose paintings and short stories delineate Baroda's landscape as dotted with domestic as well as public spaces where men unite. The image of his long-term companion, Vallavbhai, a retired building contractor, recurs in several paintings.

The gay community in Baroda could claim to be a community for purposes of marriage, and marriages conducted within this community could claim legal validity.

A Hundred Million Authorities

Another way same-sex marriages performed by customary rites can claim legal status is with regard to the fact that Hindu law always prioritizes local custom and practice over general theory (see chapter 2). Mahant Ram Puri, of *juna akhara*, who was also asked his opinion of same-sex marriage, at the Kumbha Mela, said, "There is a principle in all Hindu law that local always has precedence. In other words, the general rules and the general laws are always overruled by a local situation. I do not think that this is something that is decided on a theoretical level. We do not have a rule book in Hinduism. We have a hundred million authorities."[3]

This legal principle is an expression of Hinduism's assumption that every individual unit is a manifestation of universal energy or spirit. This assumption is the philosophical basis for the much-discussed notion of Hindu tolerance, the existence of which many today deny because so many Hindus are intolerant and violent. However, just as Christians' violence does not diminish Christianity's insistence on nonviolence, so also Hindus' intolerance does not detract from the importance of this Hindu principle.

According to this principle, the weddings that have been conducted by customary rites and accepted in local communities are legally valid Hindu weddings.

"Many Branches on the Tree of Life"

Many Hindu teachers, especially those who have taken *sannyas* (renunciation), and are thus outside the normative social order, are open-minded and accept everyone, but are hampered by followers who have not been able to shed many prejudices, including homophobia. This conflict may be seen as encapsulating the larger conflict in Indian society between philosophically tolerant traditions and intolerant social practice.

Jim Gilman, an American student of Swami Chinmayananda for 17 years till the Swami's death, and now a lay teacher of Vedanta philosophy outside the Chinmaya Mission, discusses this contradiction. He says that Swamiji had no problems with gay people, because, like most spiritual masters, he practiced "complete acceptance of everyone." Jim recounts that there were many gay men at one Chinmaya Mission camp, which upset a follower, who asked Swamiji, "What is your opinion of homosexuality?" Swamiji answered, "There are many branches on the tree of life. Full stop. Next question."[4]

Since there was no intensive discussion of the topic in the Mission other leaders and followers remained homophobic. This resulted in an upsurge of intolerance after Swamiji's death. Jim was an *acharya* (teacher) in the Mission, and in 1993, Swamiji asked him to take *sannyas* (vows of renunciation). Jim agreed and took a vow of celibacy, but Swamiji died a few months later, and his successor, Swami Tejomayananda, refused to give Jim *sannyas*. A while later, even though Jim remained celibate, he was asked to leave the Mission, because some followers were upset that he was gay.

Indian communities and groupings of different types, ranging from churches and ashrams to political parties, are losing the energies of some of their most devoted members because they are intolerant of those members' sexualities. In chapter 4, I related the story of a female couple who committed suicide after being expelled from an Ashram in south India. This type of suicide is not uncommon. A couple of years ago, an Eastern European devotee named Damodara hanged himself in a Vaishnava ashram in the United States, after an Indian ashram had cancelled his trip to India when they found out he was gay. Gaudiya Vaishnava monk Bhakti Tirtha Swami, wrote a soul-searching letter: "Recently, I have been making so much more effort in trying to open up my heart to be more available in understanding and serving all Vaishnavas . . . After hearing of Damodara's suicide . . . I must say that I have seen the light. . . . Because I was always brought up to be respectful and to try to understand all people, I did not really allow myself to go deep in trying to understand the third sex. I figured that this is necessary for those who were insensitive, arrogant and fundamentalist. . . . But perhaps worse than such bigots are those like myself who have a little understanding and think we have a lot . . . our own prejudice can easily cause us to see and not to see, *pasyannapi na pasyati*."[5]

"A Debate is Required"

Today, some spiritual, religious, cultural, and political leaders across the globe bless and support same-sex unions, even while others condemn it. As discussed in chapters 3 and 4, some Shaiva and Vaishnava priests, including Srinivasa Raghavachariar, head of the Srirangam Math in 1977, as well as several Gurus from different traditions have opined that there is no intrinsic difference between same-sex and cross-sex desire. Yet homophobia continues to be dominant among most middle-class Hindus and in many Hindu communities.

Indian debates on sexuality were stifled to some extent during the Victorian period, which in spirit extended in India through most of the twentieth century. In the last few years, however, those debates appear to be reviving.

Swami Bodhananda Saraswati is a Vedanta master in the Saraswati lineage, who took *sannyas* from Swami Chinmayananda, and is the founder of the Sambodh Foundation, which has many branches worldwide. When I discussed the matter with him, Swami Bodhananda emphasized the need for debate: "There is no official position in Hinduism. From a spiritual or even ethical standpoint, we don't find anything wrong in it. We don't look at the body or the memories; we always look at everyone as spirit. . . . It's a Christian idea that it is wrong. From a Hindu standpoint, there is nothing wrong because there is nothing against it in scripture. . . . Different priests may or may not perform same-sex weddings—it is their individual choice because there is no one position or one head of Hinduism. I am not opposed to relationships or unions—people's karma brings them together. Sexual attraction is not under your control. . . . Everyone comes into the world with their own set of needs and talents, and tries to fulfill their needs and express their talents in relationship with others. The problems are the same, whether in a gay marriage or a heterosexual marriage."

While Swamiji asserts that no objections to same-sex unions arise from Hindu philosophy, he notes that Hindu social opinion is often very negative: "We have to face this issue now. . . . I am sure spiritual persons will have no objection when two people come together. But it is a social stigma. . . . So what is required is a debate in society. I have not debated it enough. I have to do that. I have a lot of people confiding in me, 'I am very worried. I am gay. What should I do now?' I ask them to relax, ascertain their feelings, and not to worry; there is nothing wrong in that."[6]

Swami Bodhananda is not alone in this supportive attitude. A devotee of Gurumayi Chidvilasananda (successor of Swami Muktananda, and head of Siddha Yoga Foundation since 1982) reports that Gurumayi "has unabashedly come out in support of same-sex partnerships." When the state of Vermont legalized civil unions in 2000, Gurumayi responded, "Fabulous."[7]

On the other hand, several Swamis questioned at the Kumbha Mela by the *Hinduism Today* reporter were opposed to gay marriages. They did not cite any specific tradition or text, but simply declared their opposition. Swami Pragyanand of *avahan akhara* said, "Gay marriages do not fit with our culture and heritage. All those people who are raising demand for approving such marriages in India are doing so under the influence of the West. . . . we do not even discuss it." But this position was questioned from within the same tradition, for Mahant Madhusudan Giri, popularly known as Nepali Baba, who is also of *avahan akhara*, said, "Today people are even changing their sexes. They want a lot of freedom, and this freedom is available to them. So if they choose to live in a particular way out of the consent of two grown up people, how can we stop them?" Swami Avdheshananda of the *juna akhara* called the idea "unnatural, uncommon and unusual," even though Mahant Ram Puri, of the same *akhara*, had given a contrary opinion, as mentioned earlier.[8]

Tripurari Swami, a monk in the Gaudiya Vaishnava tradition, states, "My opinion regarding gay and lesbian devotees is that they should be honored in terms of their devotion and spiritual progress. They should cultivate spiritual life from either a celibate status, or in something analogous to a heterosexual monogamous situation. Gay and lesbian people have always been a part of society from Vedic times. . . . Although my Guru Maharaja [Srila Prabhupada] frowned on homosexuality in general, he was also very practical, flexible, and compassionate. One of his earliest disciples was a gay man who once related how he had ultimately discussed his sexual orientation with Srila Prabhupada. He said that at that point Srila Prabhupada said, 'Then just find a nice boy, stay with him and practice Krishna consciousness.' . . . I believe that Hinduism originally held a much more broadminded view on sexuality than many of its expressions do today."[9]

Discussion in Other Religions

Discussion of same-sex marriage in Christianity and Judaism has been going on for years in the West. Other religions too have begun discussing this issue. In a youth forum on a website run by a team of pious Sikhs, several young people ask what Sikhism says about homosexuality. The respondents note that the Sikh holy book, *Guru Granth Saheb*, says nothing about homosexuality, and that, like other desires, this desire is inborn in some people due to their karma or attachments from former lives. Enquirers are also told that all should be accepted since God is in all, and that everyone is born for spiritual experience. Although Sikh high priests in Canada opposed the Canadian government's decision to legalize same-sex marriage, the World Sikh organization supports the decision. Mr. Inderjeet Singh Bal says, "same-sex marriage is the right of a minority community, which must be conceded." Mr. Navdeep Singh Bains, the youngest member of Canada's House of Commons, agrees that same-sex marriage does not violate any religious belief as religious organizations can independently decide whether or not they want to conduct such marriages. He stated that he, like most other members of parliament would vote in favor of same-sex civil marriage as it is a question of civil rights of a minority community.[10]

In answer to a question from a Sikh in Ontario, Canada, where same-sex marriages are now valid, the respondent says that since most Sikhs, like most Indians, suffer from homophobia, it will take many years of soul-searching within the community before the question of religious blessing for same-sex marriage can be addressed.[11] In the general Discussion Forum on this website, homosexuality evokes many views, positive and negative.

I have pointed out that ancient and medieval Hindu texts freely discussed and debated many issues, including that of same-sex attachments. Medieval Islam also had a tradition of vigorous debate (*ijtihad*). Today, fundamentalists are trying to shut down these debates. Some Muslims are trying to revive *ijtihad* internationally, discussing various topics, including same-sex marriage, and using the internet among other means.[12]

From Rites to Rights

Many couples worldwide go through the rites of marriage in private. A south Indian gay male friend recounted to me in an email message how he and his male partner of six years visited two temples soon after they started living together, had special *pujas* done, and "took vows that I consider as our 'sealing' of the relationship in 'matrimony.' And ever since we met we have been faithful to each other."

In 1991, a woman from Delhi, anonymously writing about her relationship with another woman, described how they considered getting married but were afraid of social reactions, so decided to marry secretly: "One evening we went to a *mandir* [temple] and got the blessings of the deity. When we returned to the hostel, she applied '*sindhoor*' [vermilion, sign of marriage] on my '*mang*' [hair parting]. It was the happiest day for us. We never informed anyone else about our mutual pact." When they broke up a year later, she attempted suicide; she writes, "Why can't two girls get married? . . . the most traumatic thing is that the world is neither aware of our 'marriage' or of the end. I had to face the pain more or less by myself."[13]

Czech-born tennis player Martina Navratilova, now an American citizen, narrates in her 1985 autobiography how she and her partner, Judy Nelson, privately exchanged rings and vows in an empty church in Australia. Paradoxically, this union was treated as equivalent to a marriage only when it broke up, and Nelson sued for and obtained financial support from Navratilova.

Some couples, as this book shows, perform such rites publicly. In India, most of these couples have not termed themselves gay or lesbian, and have not been aware of other such marriages. None of these couples sought media attention or public attention beyond their local community.

Sheela and Sree Nandu constitute a partial exception, since they decided to talk back to the media, after a local tabloid defamed them. More recently, couples who identify themselves as gay or lesbian and are aware of other same-sex marriages, as well as of LGBT movements, have begun to marry.

In the West, many churches now conduct same-sex marriages and many more bless same-sex unions. Many rabbis in the Reform branch of Judaism perform same-sex weddings. Some couples who married in private years ago now renew their vows in public. Louie Crew and Ernest Clay, a white man and a black man who married in private, reading from the Book of Common Prayer, in 1974 are an example. Living together openly since then in the rural American South, they faced constant hostility and harassment.[14] Twenty-five years later, in 1999, they renewed their vows in the Episcopal Church, and were blessed by a bishop in the presence of a large congregation. What was unthinkable fifty years ago is now so common that many department stores register same-sex couples for gifts, and bridal businesses offer them wedding merchandise.

A shift in public consciousness and social practice occurs gradually, as these couples begin to ask other employers, insurance companies, the state, and other institutions to recognize their marriages. In the United States, many major corporations, universities, city governments, and other employers now recognize same-sex domestic partnerships for purposes of health insurance and other benefits, even though the federal government and the military do not. Numerous facilities, such as health clubs, gyms, and museums accept same-sex couples at family membership rates. More than 500 newspapers nationwide now publish announcements of same-sex unions along with announcements of cross-sex weddings. In 2002, the *New York Times* joined this group.

These changes are the result of continuous struggle and legal battles by many individuals, couples, and groups. Those battles have only recently made international headlines, but have been waged for decades. In 1975, for example, a clerk in Boulder, Colorado, issued marriage licenses to six same-sex couples. Among them were Richard Adams, an American, and, Anthony Sullivan, an Australian. They fought a nine-year battle to have Adams recognized as a spouse for immigration purposes, but in 1985, a federal judge ruled against them. Now aged 62 and 57, they live quietly in the United States, where Sullivan is an illegal immigrant.

Progress Toward Legal Recognition Worldwide

Same-sex marriages are now legally recognized in two countries; civil unions are recognized in many more. France instituted civil unions, cross-sex and same-sex, in 1988, Denmark in 1989, Norway in 1993, Sweden in 1994,

Iceland in 1996, Germany in 2001, Switzerland in 2002. The Netherlands legalized civil union in 1998 and same-sex marriage in 1998; Belgium legalized same-sex marriage in 2003; Spain, Canada, and the other Scandinavian countries are expected to legalize marriage soon.

Brazil allows partners in a "stable union" to inherit pension and social security benefits. Similar provisions are being instituted in the United Kingdom, and being debated in Chile, Colombia, and Mexico. The city of Buenos Aires in Argentina legalized civil unions in 2003. Some countries allow a person to sponsor his/her same-sex partner for immigration: Australia, Belgium, Brazil, Canada, Denmark, Finland, France, Germany, Iceland, Israel, Netherlands, and New Zealand. Several states in Canada have legalized same-sex marriage.

In the United States, changes are occurring piecemeal, and the battle is being fought state by state as well as at the federal level. In 1993, the Hawaii Supreme Court ruled that banning same-sex marriage violated the state Constitution's equal rights clause. In response, the U.S. federal government in 1996 passed an absurdly named Defense of Marriage Act, which defines marriage as heterosexual, and says no state can be forced to recognize same-sex marriages that take place in another state. Several states passed similar Acts. These laws violate the U.S. Constitution's requirement that states respect each other's legal acts. Hawaii subsequently amended its Constitution to prevent same-sex marriages, but the states of California, Hawaii, and New Jersey recognize same-sex domestic partnerships. The state of Vermont legalized civil union in 2000, and Massachusetts legalized same-sex marriage in 2004. Couples in these five states get some benefits from the state governments, but are deprived of benefits conferred on heterosexual married couples by the federal government. Some states are now amending their Constitutions to institutionalize inequality, and a movement is afoot to amend the U.S. Constitution in the same way.

Following the Massachusetts decision, officials across the country, from San Francisco to New York and Oregon to New Mexico, engaged in acts of civil disobedience, marrying thousands of same-sex couples. These marriages the state does not recognize are equally important in terms of social impact. Watching the 4.037 San Francisco marriages on TV in 2004, Cambodian King Norodom Sihanouk, aged 81, father of fourteen children, argued on his website that as a democracy, Cambodia should legalize gay marriage; he remarked that "God loves a diversity of tastes and colors in all species, humans, animals, vegetables."[15]

India's International Same-Sex Marriages

The Indian subcontinent has engaged in global connections for millennia. The recent wave of Indian emigration is only one in many such waves over the centuries. International marriages are one of the consequences of immigration. In most cases, the couples live abroad, but there are also several foreign spouses who settle in India. In either case, the Indian partners usually retain strong connections with families in India, who have responded in a variety of ways to these marriages. Some of these international marriages are same-sex marriages.

In January 1993, Aditya Advani married his partner, Michael Tarr, an American, in the Advani home in New Delhi (photo 10.1). Aditya had remarked to his mother, Kanta Advani, who lives in Delhi, that no one would ever come to his wedding, and she responded, "Why not? We could have a ceremony for you and Michael."[16] Aditya then proposed to Michael, and Aditya's parents issued invitations to friends and relatives. Aditya's spiritual teacher, who believes that being homosexual or heterosexual is immaterial to spiritual progress, agreed to marry them. He performed a *puja* of Lord Ayyappa, son of Shiva and Vishnu. In a private conversation with me, the officiant remarked, "My attitude was, 'Here are two people who want to live together so it has to have the blessings of God, parents and elders. I chanted a mantra from the Upanishads for their prosperity, for their love, so that they can live together, so that their mind is clear, intellect is strong, they may have respect from everyone, enough money, all those kinds of things, they will not be disliked by others, they will be loved by others.' " The couple

Photo 10.1 Aditya's mother, Mrs. Kanta Advani (extreme left), with Aditya and Michael, during their wedding ceremony, conducted by the Advanis' spiritual preceptor with Hindu rites at the Advani family home in New Delhi, 1993.

walked round the fire, exchanged garlands, and touched the feet of elders. Many friends and relatives attended and gave their blessings.

Arvind Kumar and Ashok Jethanandani are pioneering activists in the South Asian LGBT community in the United States. Arvind is one of the founders of *Trikone* magazine. They are both engineers, who were raised in Bihar, but met in California, where they have lived together since 1986. They got married in 1996 on their tenth anniversary (see photo 10.2). Arvind's mother is an ascetic. Her name is Ma Yogashakti; she took *sannyas* in 1960, in the *Niranjani akhara*, and has ashrams in the United States and in India. She initially disapproved of Arvind's being gay, and thought he would never be happy. But eight years after he got together with Ashok, when she realized they were happy together, she suggested that they marry, and offered to perform the ceremony. The wedding took place in Toronto, with Arvind's entire family present. Arvind says, "I finally entered the family circle."[17]

Ashok's family could not be present at the wedding because it had not been planned ahead of time. Ashok's father was initially very disapproving

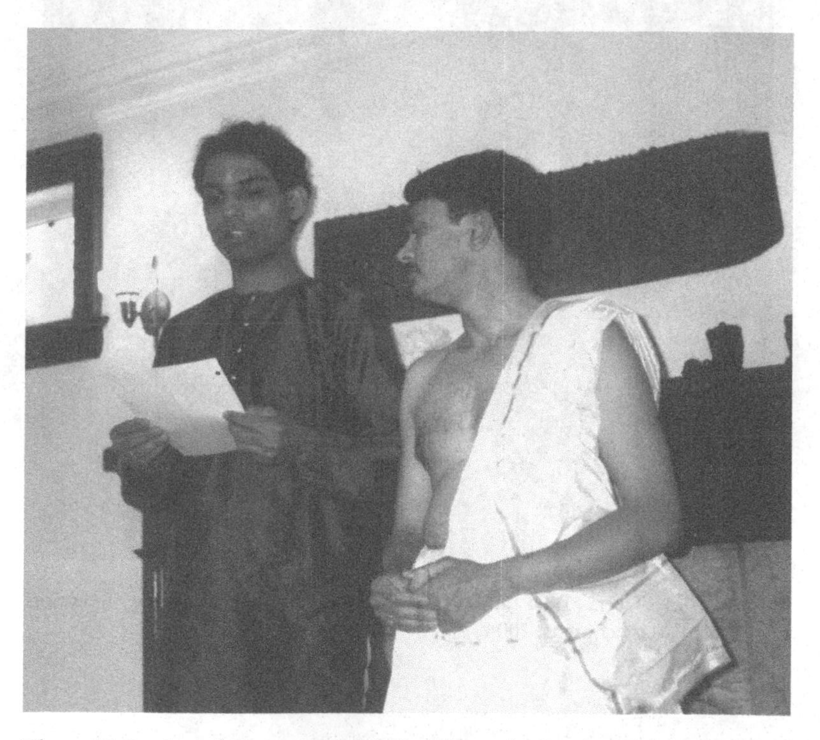

Photo 10.2 Arvind Kumar and Ashok Jethanandani exchanging vows at their wedding conducted by Arvind's mother, Ma Yogashakti, in Toronto, 1996. Photo by Anil Verma.

Photo 10.3 Geeta Patel and Kath Weston at their wedding reception, Cambridge, Massachusetts, 2004.

of his homosexuality but his mother always accepted it; over time, Ashok's father changed his attitude; both parents migrated to the United States and now live with Ashok and Arvind.

In June 2000, Mona Bachmann and I married in New York, with a Hindu ceremony (*saptapadi* and *jaimala*), and a Jewish ceremony conducted by a rabbi; we held a reception at India International Center, Delhi, the following month. In June 2002, Vegavahini Subramaniam and Vaijayanthimala Nagarajan (generally known as Vega and Mala) married, in a Hindu ceremony performed by a Shaiva *pandit*, in Seattle, followed by a party in the Seattle Aquarium. They later joined a lawsuit challenging the state of Washington's non-recognition of same-sex marriage.

In June 2003, eminent geologist Rustam Kothavala, now 70, who hails from Bangalore, married his partner of 30 years, sociologist Toby Marotta, under Vermont's civil union law.[18] In June 2004, Geeta Patel, professor at Wellesley College, married her partner of eleven years, Kath Weston, director of women's studies at Harvard, in a Hindu and Buddhist ceremony, conducted by a Buddhist nun, at the Cambridge Zen Center; this marriage is legally recognized under Massachusetts' new law (see photo 10.3).

The 2002 union of Indian fashion designer Wendell Rodricks, a Goanese Catholic, with Jerome Marrell, his partner of 20 years, was widely reported by Indian newspapers as the first legal same-sex union on Indian soil, because Jerome is French, and the union was performed by an official of the French Consulate, according to French law. The ceremony, on December 26, 2002, at their home in Wendell's ancestral village, Bardes, Goa, was attended by family and friends, including celebrities from the fashion and film industries, who flew in from all over the country. Jerome is a chef, and Wendell was also a chef before he became a designer; the couple has now opened a French restaurant in Goa. Wendell, who had not sought press publicity, noted that the news received "an overwhelming positive reaction."[19]

Changing and Unchanging Realities

The Canadian Oxford Dictionary has altered its definition of marriage to "the legal or religious union of two people." In addition to the deliberately non-gendered noun "people," this definition recognizes that marital unions not validated by some governments are nevertheless marriages. As social reality changes, laws and governments will also be compelled to change.

Marriage is the type of union most easily and universally recognized. A couple from one country who get into difficulties in another country are most likely to be understood if they say they are married, rather than trying to explain that they are civilly united, domestic partners or reciprocal beneficiaries. The most important reason to grant same-sex couples the right to a legal civil marriage is that many of us want this right, and our having it will do no harm to anyone else. It will end social and financial discrimination, encourage stability, responsibility and mutual accountability between partners, and bring the same benefits to society at large that cross-sex marriage is thought to bring. It will enable our families and communities to understand our unions as fully equal to heterosexual unions, and to embrace and bless them in public.

Until such changes occur, couples will continue to battle for their lives. I began this book with the story of Mallika and Lalithambika, who attempted suicide in 1980. As this book goes to press, twenty-five years later, reports appear nearly every week of couples eloping, being arrested or forcibly separated, and committing suicide. The latest tragedy is that of Neetu Singh, aged 20, and Ranu Mishra, aged 22, in Lucknow, who attempted joint suicide, after Ranu's parents arranged her marriage to a man. Both consumed poison; Neetu died, and Ranu is in critical condition in hospital. Ranu, who had the wedding *sindoor* in her hair parting, told police she had taken Neetu as her husband, and that they loved each other and could not live apart.[20]

At the 2002 wedding of Mala and Vega, Vega's father read a poem in Tamil that he had composed. In the poem, he quoted the ancient Tamil poet Valluvan, to explicate his own support of his daughter's wedding.

> Inquiring as to what is love, realizing that the best words are those of the great Tamil seer,
> I found the book, opened and read it.
> Valluvan says . . . .
> "The seat of life is love; anyone who does not have it is only a mass of bones encased by skin."
> Hence . . . on this auspicious day when you are avowing the
> Bonds between you, in the presence of this august audience, as a father, I rise to greet you and give my blessings,
> As two bodies with one soul, let the love you bear each other today . . . grow forever and ever.[21]

Appendix: Note on Theory and Methodology

This book is not a history, exhaustive or otherwise, of same-sex unions. It traces some premodern textual precedents of such unions, to show that contemporary same-sex marriages have antecedents in traditional narratives. Although these narratives are separated by space, time, genre, and context, some patterns of representation do emerge.

I do not claim that the Indian female couples who married over the last two decades knew about these precedents. It is unlikely that they did, with the possible exception of the Sikhandin story. It is important, however, that the idea of same-sex union was, to the couples and their families, both thinkable and speakable.

This intelligibility partly results from the fact that same-sex union was debated in the past, and, despite many discontinuities, continues to be discussed today. Hindu philosophical concepts about gender, love, sex, the spirit, rebirth, and change, are widely understood and cited, even among the illiterate and semiliterate.

This is not a book about the writings of lesbian, gay, bisexual, or transgendered writers. I do not think that people's ideas are fully reducible to or explicable by their biographies. In my view, the genders, sexual practices, sexualities, castes, and classes of the authors of texts I examine are not of decisive significance. First, these can rarely be fully known, and, second, I am interested not in particular authors' biographies but in the ongoing discussion and circulation of a set of ideas over time. In a small way, this book joins that discussion, as do the media reports of same-sex weddings in India.

Histories of ideas about love and marriage in India have tended to ignore same-sex relationships, while histories of friendship yet remain to be written. Both feminist and nonfeminist historians focus on women's relations to men, largely in the modern era. Rajat Kanta Ray, otherwise an extraordinarily sensitive historian from whom I have learnt much, declares: "Of all interpersonal relationships, that between man and woman is characterized by the widest range of emotions, and of all such emotions, the psycho-sexual attachment known as love is the strongest."[1] His assumption is that "the psycho-sexual attachment known as love" occurs only between man and woman. He states that this is the strongest emotion without offering any more proof than novelist Balzac offers when his narrator refers to "the strongest emotion known—that of a woman for a woman."[2] In his history of love, Ray refers to same-sex relations only once, to issue a warning against some unnamed feminists' "lopsided preoccupation with homosexuality 'lesbianism' [that] may, moreover, result in the cultural value of heterosexual eroticism in the evolution of civilization being overlooked."[3]

If I am accused of such a "lopsided preoccupation," I can only plead guilty. I doubt very much, however, that paying attention to this neglected area of social reality will lead to heterosexuality's value being overlooked. Throughout this book, I place same-sex eroticism in the context of eroticism in general as well as the intersecting contexts of family, kinship, friendship, and gender. I also point out similarities between cross-sex and same-sex union, often neglected by those primarily interested in emphasizing difference.

Psychologist Sudhir Kakar similarly assumes that love and sexual intimacy occur in India only between men and women, and that the only *jodi* (couple) Indian women dream of is heterosexual. According to him, homosexuality is merely a "temptation" for men living away from "their women-folk." The "romantic longing for completion," is, he declares, "a gift solely in the power of a woman to bestow."[4] I hope to demonstrate that all of these generalizations are mistaken.

Sources and Methodology

Most of my sources are textual, ranging from canonical works to marginalized ones, and from newspaper reports to films and websites. It is not, therefore, part of my argument that same-sex relationships occur in life in exactly the way texts depict them. Like Terry Castle, I am interested not only in what same-sex unions are but also in what people say they are—their role as "rhetorical and cultural tropes."[5] What people say they are is not unconnected to what they are, although the connection is oblique rather than direct. This is especially the case with literary texts, such as those discussed in chapters 6 and 8, where generic conventions influence representation.

Feminist and subaltern studies historians often see written, especially Sanskrit, texts as "Brahmanical," elitist, and oppressive, and therefore focus on oral or folk texts in vernacular languages. In my view, not only are all texts hybrid and open to multiple readings but Indian written texts, especially sacred texts, are constantly in interplay with oral traditions, and Sanskrit texts with vernacular traditions. Written texts, such as the many versions of the *Ramayana*, are routinely recited, sung, and enacted, and many Sanskrit phrases, tags, and quotes are incorporated into modern languages. One example of how contemporaneous regional Sanskrit and vernacular versions of a narrative interact is discussed in chapter 6.

This is a work of cultural studies; my strategies are multidisciplinary although primarily drawn from literary studies, and my sources range across space and time. This is because I am interested in the ways same-sex unions are produced and received in culture today, and this analysis requires examining the many contexts in which they are simultaneously embedded. A culture's understanding of same-sex relationships is inextricable from its understanding of, among other things, love, marriage, friendship, gender, and human relationships with the divine and the nonhuman. Our worldviews draw upon ideas from different times and places that percolate through our cultures. An educated middle class urban Indian's understanding of these dimensions of life is, for example, likely to be inflected not only by practices in his/her and friends' families, but also by their representation in newspapers, magazines, television, and cinema, all of which report local, national and international news, and cultural events; and also by film music, pop music, Hindu devotional music and poetry, the *ghazal*, popular fiction, and fragments of the Bible and of Shakespeare.

With regard to journalistic reports of same-sex weddings, although reporters' and readers' biases do inflect them, these reports nevertheless constitute evidence that the rites of marriage have been available to some same-sex couples in India over the last two decades, outside the purview of any organized movement for same-sex marriage. That the social right to marry has not been equally available to all is tragically clear from the love suicides that have occurred in these same years.

Categories and Identities

Discussion of same-sex relations today is fraught with questions of terminology. Several writers object to identity categories such as "lesbian," "gay," and "homosexual" being used in non-Western contexts, arguing that these are neocolonial because they were (a) coined in the modern West; (b) exported to countries like India under conditions of colonialism, and are now being transmitted through postcolonial neo-imperialist globalization; (c) have no exact counterparts in non-Western languages; and (d) do not match non-Western social reality, where sexual behavior, not identity, is classified and punished.

I disagree. First, Foucault and his followers' claim that sexuality-based identity categories were invented by nineteenth-century Euro-American sexologists has been repeatedly disproved by historians, both for the West and for India, but still unaccountably persists, much as the idea that printing was invented in fourteenth-century Europe persists despite evidence that it was invented in ancient China.[6] In chapter 1, I discuss one counter-example—the *Kamasutra*, which categorizes men who are attracted to men as a third nature, and notes that while some are masculine-appearing, others are feminine-appearing, and all incline to certain professions like hairdressing, massage, and flower selling.

Second, exchange and circulation of goods, texts, and ideas between Asian and European cultures has been going on for centuries. In periods when this exchange is more intense, people tend to think that it is a new phenomenon that has never occurred before. Some Renaissance thinkers made this claim, but recent research on the Silk Road shows that such exchange was frequent in antiquity. Modern technology has speeded up the exchange and made some types of information available to larger groups of people. But, as I have argued elsewhere, the exchange of ideas is not and never has been entirely unequal or one-way, even under colonialism.[7] Even were one to accept the highly debatable claim that exchange of goods and services is entirely unequal between so-called East and so-called West, exchange of ideas is not reducible to economic exchange. If terms like "gay" are now widely used in India, references to the "third sex," for which at least one source is the *Kamasutra*, now abound in Euro-America, even in popular culture.

Nor does globalization mean that today's mobile educated populations are more citizens of the world than were their ancestors. The proliferation of information and literacy is accompanied by a provincialization, so much so that fewer educated people are now culturally literate about the past of even their own society, let alone any other. For example, more Euro-Americans today have heard of the *Kamasutra* than was the case in the nineteenth century, but few are aware of the discussion of same-sex desire in Plato's *Symposium*, which was known to every educated man up to the nineteenth century, or of the love between Achilles and Patroclus or Orestes and Pylades, with which every schoolboy in Renaissance England was familiar. Up to the nineteenth century, educated people knew several languages and uneducated people often were bilingual or trilingual; increasingly today, even highly educated people can read only one language.

Third, English has long been an Indian language, and India is now the third-largest producer of English books in the world, after the United Kingdom and the United States. Words like "lesbian" and "homosexual" appear in Hindi and Urdu texts in the early twentieth century, and the Oscar Wilde and *Well of Loneliness* trials (1895; 1928) were widely reported and debated in India; conversely, nineteenth-century European translators of Sanskrit texts made Indian categories of thought about sexuality available in Euro-America.

The attempt to replace terms like "gay" with supposedly indigenous and authentic terms in Indian languages is futile, unless we also advocate that all Indians eschew the English language, a misguided ahistorical ideal rendered increasingly impossible by ever-growing numbers of internationally mobile Indians. British journalist Jeremy Seabrook, ignoring all scholarly evidence to the contrary, insists that words in Indian languages (he cites the modern oral regional slang *kothi* and *giriya* for men, ignoring women, but does not mention the many terms attested in texts and documented by scholars) refer only to sexual behavior, not identity, and that, consequently, Indians today who identify as gay, lesbian or homosexual are the unenlightened descendants of colonial officials, and are "striking violently against the multiple competing aspects of the human person in traditional societies."[8] This formulation reifies India as "traditional" versus the modern West, and assumes that Western humans do not have multiple competing aspects.

In my view, such objections are in part constructed by homophobia. No comparable objections are raised to identities like "Professor," "Dean," or "Doctor" in non-Western contexts, even though these academic identities evolved in universities in European Christendom. Nor does the word "child" have an exact counterpart in Sanskritic languages. Does this mean that young Indian people should never be identified as children in discussion of modern or premodern Indian texts?

Fourth, terms like "gay" and "lesbian" do match Indian reality, since many Indians now identify as gay and lesbian and some have done so for several decades. Nor are sexual acts alone punished—some people are punished merely for claiming an identity. For example, when seven schoolgirls in Thiruvananthapuram, Kerala, were expelled in 1992 for forming a Martina club, they were penalized for declaring an identity, not sexual behavior.[9]

Some activists and theorists advocate the use of MSM (men who have sex with men) as an identity category more suitable for Indian men, because it focuses not on identity but on behavior. This argument is self-contradictory because "MSM" was coined by health workers in the United States with regard to American men, not as an identity category but as a medical term for use in the context of HIV prevention efforts. Almost no English speaking Indian man who is reluctant to identify as gay or bisexual would find it easier to socially identify himself as a "man who has sex with men." I further discuss some of these issues in chapter 7.

Identity to Relationship: Shifting the Focus

The identity versus behavior debate neglects emotional and relational contexts. In this book, my focus is not on identity or behavior but on relationships. Both in the West and in India, gay people's narratives of their recognition of their gayness show that individuals generally construct their sexual identities in the context of attraction to at least one other person.[10] The texts I examine demand a shift from an inward-turning focus on individual sexual identity to an outward-looking focus on relationship and kinship, as all these texts represent desire and love in the context of relationship.

As my focus is on same-sex marriage, not on individual sexual or gender identities, I discuss transgender and transsexual identities of various kinds in the context of same-sex unions. When discussing texts, I generally use the terms the texts use, and I also refer to relationships as "cross-sex" or "same-sex" rather than as heterosexual, homosexual or gay. I use terms such as "gay" and "lesbian" when discussing individuals today who identify in these ways. In these contexts, I also sometimes use "gay" to include both men and women.

Another theory currently dominant in the academy is that same-sex relations between non-Western males are circumscribed by active–passive, top–bottom, older–younger hierarchies, in ways that Western ones are not, and therefore are about power rather than love or affect. This supposedly clear-cut distinction between hierarchical models of relationship in the past and in Asia, as opposed to egalitarian models in the modern West does not correspond to reality. Active-passive, older–younger models of same-sex relationship are still widely prevalent in the West, while models based on likeness between lovers are attested in premodern texts, both Asian and Western, and also in Asia today.[11] One needs to distinguish between the normative ideal, to which people may claim to conform, and actual behavior. In interviews and on the record, men may claim to be exclusively active or passive, while in private conversation, they may reveal much more flexibility. Also, the active–passive model assumes penetrative anal sex as the dominant form of sex, but this is not proven to be the case. Some Indian gay men have told me that oral sex is very common, especially in quick encounters, and one long-term couple told me that they engage only in mutual manual sex with each other. Much more rigorous research needs to be done on sexual relationships before claims of any kind can be established.

Furthermore, active-passive and older-younger hierarchies are dominant in cross-sex relationships worldwide, yet the term "marriage" is used to encompass cross-sex unions of all kinds—egalitarian or non-egalitarian, chosen or arranged, lifelong or serial, monogamous or polygamous. That same-sex unions are also structured in a variety of ways, in different societies, should not preclude their being interpreted as conjugal.

This book carries further Kidwai's and my exploration of terms used in premodern Indian texts to identify and categorize same-sex relationships and the persons engaged in them as well as those habitually given to them. I analyze these terms and place them in their literary and cultural contexts. When literary texts, such as *rekhti* poems, use terms for certain relationships (such as a woman's female lover and their amorous relations), and these are also documented in other sources such as dictionaries, they indicate an awareness of such relationships in some strata of contemporaneous society.

To take just one example, the term *chapti*, and its variant *chapat baz* (player of *chapti*) used for lesbians and lesbian activity in Urdu, including *rekhti* poetry (see chapter 8), is attested in several sources. Not just Urdu speakers but also Hindi scholars were aware of this term. As late as 1964, Devdutta Sastri, in his Hindi translation of the *Kamasutra*, quotes a line from *rekhti* to comment on Yashodhara's twelfth-century commentary on the *Kamasutra*. Sastri writes in Hindi: "Analyzing the mutual union of men, Yashodhara mentions women also mutually engaging in vaginal rubbing. These days, this act by women is called *chapti*, '*Aao sakhi, chapti khelen, baithey se begar bhali*' Come girlfriend, let us play *chapti*, better to labor without payment than to sit idle."[12]

The "Hindu" and the Indic

The term "Hindu," a coinage by West Asian Muslims to refer to the non-Muslim inhabitants of the region they called "Hindustan," later came to be used as a religious identity category. I focus on philosophical traditions and social practices that have come to be called "Hindu," first, because most of the weddings and suicides I examine involve Hindus and Hindu wedding rites, and, second, because I follow thinkers like Ashis Nandy and Rajat Kanta Ray in the view that Indic cultures are permeated by Hindu traditions and practices, which influence both non-Hindu Indians and also nonreligious Indians.[13]

One major difference between South Asian and Southeast Asian Islam on the one hand and West Asian Islam on the other is that the former is shaped by its interface with Hindu patterns of thought and practice. These constantly changing patterns are also permeated by other cultural influences, such as the Christian, the Marxist, and so on. However, Hindus constitute a large majority of the Indian population, and, to that extent Hindu categories of thought are dominant, just as Christian categories of thought are dominant in Euro-American cultures. Rebirth is an example of such a category, and I am interested in how these categories affect ideas of same-sex attachment.

Since some thinkers claim that Hindus are too diverse to be treated as one category, I should mention that despite the many differences among Hindu communities, there are enough recognizable and continuous patterns of thought, feeling, and practice among them to make the category meaningful, just as there is enough in common between various Christian denominations to render that category meaningful.

India and the West: Exceeding the Postcolonial

Comparative cultural studies are suspect in the academy today, and comparing Western and non-Western cultures is taboo, following Said's denunciation of Orientalist scholars who made such comparisons. Comparisons are, however, inevitable, especially with regard to the debate around same-sex relationships, which is today, for better or worse, an international one.

One way of addressing these connections is through the anxieties of the postcolonial; another way, which I adopt in this book, is through the recognition that ideas and texts have continuously circulated between Indian and Western cultures for centuries before colonialism, and have continued to do so during and after colonialism.[14] Postcoloniality is just one, and not the most important of, the historical factors that shape modern Indian cultures. Precolonial texts, written, visual and iconic, are still read, recited, enacted, worshiped, and rewritten throughout Indian society, thus shaping and being shaped by the lives and imaginations of modern Indians. Thinkers went global much before multinational companies did so. Following in their footsteps, I hope that my explorations of Indian ideas of same-sex union, love, marriage, friendship, gender, reproduction, and sanctity, born of countless circulations, may prove fruitful in contexts other than the texts where I find them.

India today, somewhat like premodern Europe, comprises a wide variety of marriage practices. This variety of practices has not yet been eliminated by the modern state, although that process is underway and may soon be complete. Looking at same-sex marriage in the Indian context allows us to view it in the perspective of a diverse terrain rather than the relatively more uniform marital landscape prevailing in Western democracies.

Public discussion of same-sex marriage in India today, largely sparked off by the female-female weddings reported in newspapers over the last two and a half decades, functions as a fascinating counterpoint to the debate in the West, and vice versa. Comparing the two debates allows us to examine how modern interpretations of different religious traditions affect ideas of sexuality. I focus mainly on Christianity and Hinduism, the traditions I know best. This is also my reason for choosing India rather than other Asian civilizations, such as China or Japan.

The term "West," like "India," and like most such terms for civilizations rather than nations, is permeable and fluid. However, it is widely used and understood, especially in India, and in this book refers to Western Europe, the United States, and Canada. The high-profile battle being waged over same-sex marriage in the United States is getting more attention in the Indian and international press than much longer-standing battles in European countries and Canada. This should not lead us to forget, though, that struggles in other countries have been equally intense—the Netherlands' legalization of same-sex marriage was the culmination of two decades of struggle.

As a democracy, India faces the same questions regarding civil rights of minorities, including gay people, faced by Western democracies. In its 1991 report, *Less than Gay: A Citizens' Report on the Status of Homosexuality in India*, the ABVA (AIDS Bhedbhav Virodhi Andolan) formulated a "charter of demands," of which the fifteenth is that the government of India "amend the Special Marriages Act to allow for marriages between people of the same sex."[15] Since then, several gay rights and human rights organizations have reiterated the demand.[16] As this book documents, some couples have sought recognition for their marriages from local governmental authorities. It is only a matter of time before the demand is taken to court.

The comparison is also interesting because of the Indian stereotype of the West, as a homogenized wildly permissive society where the family has broken down, and the Western stereotype of India as a highly traditional society in which all women submit to oppression, and all families are exactly the same—patriarchal, religious, and heterosexual. But, as Ashis Nandy remarks, "the greater Sanskritic culture, while institutionally one of the most rigid, has always been ideologically one of the most tolerant."[17] I would add that social reality is not uniformly rigid either. A wide variety of living arrangements coexist in India. Perhaps the most surprising I have encountered was in the late 1970s in a small town in the Himalayas, where I met a family consisting of two men and a woman, in their fifties, all unrelated and unmarried to one another. They were in every respect regular middle-class people, except for their unusual living arrangement, which they had maintained since their college days.

Conversely, many Indians congratulate themselves on the low divorce rate in India, as compared to Western countries. What they forget is that statistics maintained by the Indian government regarding divorce are completely unrealistic. Most Indians do not register their marriages with the government, and many of those who divorce and remarry also do so without informing the government. Many incompatible couples separate and live apart all their lives, without divorcing or remarrying. The rate of breakdown of families is, therefore, far higher than government statistics suggest.

Mortality rates, especially maternal mortality rates, must also be taken into account when examining marriage. Lawrence Stone has shown that before the nineteenth century high mortality rates in Europe resulted in many people, especially men, marrying more than once, and when modern medicine lowered mortality rates, divorce became more common. He remarks, "modern divorce is little more than a functional substitute for death."[18]

Idealizing the Past?

When presenting this and related work, especially in South Asianist academic forums, I have been warned against positing a narrative of decline and idealizing or romanticizing the past. While I do argue that homophobia in its present form is a modern development in India, I also see modernity as the site of new opportunities and spaces, among them greater economic and social independence and mobility for many people, especially women. This new social reality has facilitated the same-sex weddings of the last two decades. Therefore, this book does not narrate a story of decline.

As far as idealization is concerned, I hope that my stance is one of critical appreciation and my strategy one of rigorous close reading. To say anything positive about premodern religious texts, especially Hindu texts, is today often seen as retrogressive. When I presented some of the materials from *Same-Sex Love in India* at a feminist retreat in India a couple of years ago, a leading Indian feminist historian questioned the need to research the history of same-sex desire, arguing that such research could further the agenda of the Hindu right. In a similar vein, some Indian feminists expressed shock at the existence of a *puja* room in a lesbian feminist's house in Delhi, and surprise at my learning Sanskrit (although not at my learning Urdu), and at the Hindu part of my wedding ceremony.

While aware of the many oppressive dimensions of religious texts, I am not willing to negate the joy or awe with which their imaginative abundance and wisdom have often surprised me. Hindu and Christian traditions are not the property of right-wing forces any more than marriage, kinship, and family are the property of heterosexuals. The best antidote to ignorance and literalism of all kinds is the exploration and discussion of all types of evidence, sources, and ideas.

Translation

All translations from Hindi and Urdu are by me, unless otherwise indicated. Translations from Sanskrit that are by me are indicated in their respective endnotes. When English translations are easily available elsewhere, I refer the reader to them, and give priority to texts not formerly translated. Space constraints prevent me from always transliterating the original, although I do so when necessary.

I use first names to refer to the women whose marriages/joint suicides have been reported in the Indian press over the last two decades. This is because Indian newspaper reports that I quote have consistently referred to them in this way, often not providing surnames. Surnames are a relatively recent phenomenon in India (introduced by the British) and are still not used by large sections of the population.

Notes

Chapter 1 Introduction

1. Quoted in Frank Rich, "The Joy of Gay Marriage," *New York Times*, February 24, 2004.
2. Robert Wintemute and Mads Andenaes, *Legal Recognition of Same-Sex Partnerships: A Study of National, European and International Law* (Portland, OR: Hart Publishing, 2001).
3. See Evan Wolfson, *Why Marriage Matters* (New York: Simon & Schuster, 2004).
4. All citations from the Bible refer to the Authorized King James Version, with Apocrypha (Oxford: Oxford University Press, 1997). All citations from the *Mahabharata* refer to *The Mahabharata of Krishna Dwaipayana Vyasa*, translated Kisari Mohan Ganguly (1883–1896; New Delhi: Munshiram Manoharlal, 1970, 3rd edn. 1973).
5. See Ruth Vanita and Saleem Kidwai, *Same-Sex Love in India: Readings from Literature and History* (New York: St. Martin's, 2000; New Delhi: Macmillan, 2002), hereafter cited as *SSLI*; Louis Crompton, *Homosexuality and Civilization* (Belknap Press, 2003); Bret Hinsch, *Passions of the Cut Sleeve: The Male Homosexual Tradition in China* (University of California Press, 1990); Gary P. Leupp, *Male Colors: The Construction of Homosexuality in Tokugawa Japan* (University of California Press, rept. 1997); Will Roscoe, *Changing Ones: Third and Fourth Genders in Native North America* (New York: Palgrave, 2000); Stephen O. Murray et al., *Boy-Wives and Female Husbands: Studies of African Homosexualities* (New York: St. Martin's, 1998).
6. Greg Reeder, "Same-Sex Desire, Conjugal Constructs, and the Tomb of Niankhkhnum and Khnumhotep," *World Archaeology* 32:2, October 2000. See also www. Egyptology.com/reeder/.
7. John Boswell, *Same-Sex Unions in Premodern Europe* (New York: Random House, 1994), especially the chapter, " 'A Friend Inspired by God': Same-Sex Unions in the Greco-Roman World."
8. See Helena Whitbread, ed., *"I Know My Own Heart": The Diaries of Anne Lister, 1791–1840* (New York: New York University Press, 1992), and Alan Bray, *The Friend* (Chicago: Chicago University Press, 2003).
9. Bruce Bagemihl, *Biological Exuberance: Animal Homosexuality and Natural Diversity* (New York: St. Martin's, 1998).

10. See Sheikh Abrar Husain, *Marriage Customs among Muslims in India* (Delhi: Sterling, 1976).
11. See Martin Ottenheimer, *Forbidden Relatives: The American Myth of Cousin Marriage* (Champaign, IL: University of Illinois Press, 1996).
12. For this history in the West, see E.J. Graff, *What is Marriage For?* (Boston: Beacon Press, 1999).
13. "Women's Suicide Pact," *The Tribune*, December 1, 1980.
14. "Legal Seal on Lesbian Marriage," *Statesman*, December 13, 2004.
15. " 'Girlfriends' in Suicide Pact," *Blitz*, July 11, 1980, p. 7.
16. See Martha Nussbaum, *Hiding from Humanity: Disgust, Shame and the Law* (Princeton, NJ: Princeton University Press, 2004).
17. Mark D. Jordan, *The Invention of Sodomy in Christian Theology* (Chicago: University of Chicago Press, 1997).
18. Vanita and Kidwai, *SSLI*, 168.
19. See Bina Fernandez and Gomathy N.B., *The Nature of Violence Faced by Lesbian Women in India* (Mumbai: Tata Institute of Social Sciences, 2003).
20. See Byrne Fone, *Homophobia: A History* (New York: Henry Holt, 2000); Michael Rocke, *Forbidden Friendships: Homosexuality and Male Culture in Renaissance Florence* (New York: Oxford University Press, 1996).
21. For prosecutions under this law, see Suparna Bhaskaran, "The Politics of Penetration," in *Queering India* ed. Ruth Vanita (New York: Routledge, 2002), 15–29.
22. See http://www.hindustantimes.com/news/printedition/221203/detFRO03.shtml.
23. Romila Thapar, *Sakuntala: Texts, Readings, Histories* (Delhi: Kali for Women, 1999), 200–201, 236–242, and Tagore's reinterpretation of Kalidasa's *Sakuntala*, ibid., 242–250.
24. Vanita and Kidwai, *SSLI*, 191–197, 200–207, 236–240, 246–252.
25. Justin Deabler, "More than the Gospel Truth" August 10, 2004, www.advocate.com.
26. J. Rajasekharan Nair, "I Believe," *Savvy*, February 2004, p. 20.
27. "Lesbians Marry in Amritsar,"*Entertainment News*, December 13, 2004.
28. "Girls on the Run," *The Week*, October 1, 2000.
29. See the website of Straight Alliance for Marriage Equality, www. sameproject.org.
30. Kath Weston, *Families We Choose: Lesbians, Gays, Kinship* (New York: Columbia University Press, 1991), 210.
31. Vanita and Kidwai, *SSLI*, 122.
32. Montaigne, "Of Friendship," in *The Complete Essays of Montaigne*, translated Donald M. Frame (Stanford: Stanford University Press, 2002), 135–144;141.
33. As early as the 1950s Jessie Bernard noted this in *Remarriage: A Study of Marriage* (New York: Dryden, 1956), 119.
34. "Jaya and Tanuja: Two Spectrums of Human Relationships," *Times of India*, May 25, 2001.
35. For a cross-caste love marriage, followed by the bride's natal family murdering her husband and in-laws, and seriously injuring her, see "Murder did them part," *The Week*, June 20, 1999, p. 14.

36. "Do Ladkiyon ki Dilchasp Shadi" (The Interesting Wedding of Two Girls), *Madhur Kathayen* 49 (October 1993).
37. *Trikone Magazine* 16:4 (October 2001), 20.
38. Vijay Kumar Rastogi, "Lesbian Couple Sparks Debate in UP," *Sify News*, April 19, 2005.
39. Details from letter written by Sheela and Sree Nandu to the Commissioner of Police, Thiruvananthapuram.
40. Martha Nussbaum, "Steerforth's Arm: Love and the Moral Point of View," *Love's Knowledge* (New York: Oxford University Press, 1990), 335–364.
41. Nussbaum argues that since we can deceive ourselves in love, suffering proves the authenticity of our own love to us. *Love's Knowledge*, 261–285.
42. Andrew K.T. Yip, *Gay Male Christian Couples: Life Stories* (Westport, CT: Praeger, 1997), 92.
43. Mel White, *Stranger at the Gate: To Be Gay and Christian in America* (Plume Books, rept. 1995); Rabbi Steven Greenberg, *Wrestling with God and Men: Homosexuality in the Jewish Tradition* (Madison: University of Wisconsin Press, 2004).
44. Quoted in Ankur Rupani, "Sexuality and Spirituality," *Trikone* 18:4 (2003), 15.
45. Isherwood, *My Guru and His Disciple* (New York: Penguin, 1980), 25, 254.
46. Ashok Row Kavi, "The Contract of Silence," in *Yaraana: Gay Writings from India* ed. Hoshang Merchant (Delhi: Penguin, 1999), 12–15.
47. www.hinduismtoday.com/archives/1994/7/1994-7-09.shtml
48. Amara Das Wilhelm, *Tritiya-Prakriti: People of the Third Sex* (Galva and Xlibris, 2003).
49. "Happy to be Gay," *Times of India*, June 17, 2004.
50. See Arvind Sharma, "Homosexuality and Hinduism," in *Homosexuality and World Religions* ed. Arlene Swidler (Valley Forge, PA: Trinity Press, 1993), 47–80; *Trikone* special issue "Hinduism and Homosexuality," 11:3 (July 1996); and my discussions in the Preface and Introductions to Ancient, Medieval Sanskritic, and Modern materials, in *SSLI*.
51. *Padma Purana* translated N.A. Deshpande (Delhi: Motilal Banarsidass, 1990), VI. 112. 23b–33a. All further citations are to this edition.
52. See *The Laws of Manu* translated Wendy Doniger with Brian K. Smith (Delhi: Penguin, 1991). In 11:174 (p. 267), the penance prescribed for a man who ejaculates in something other than a vagina is a minor one, also prescribed for "stealing articles of little value" (II.165).
53. Mark D. Jordan, *The Invention of Sodomy in Christian Theology* (Chicago: University of Chicago Press, 1997), 17.
54. Both the preceptors in the *Mahabharata*, Dronacharya and Kripacharya, are born in this way. A divine variant of this is when Shiva, interrupted during intercourse with his wife, ejaculates into the fire god Agni's hands or mouth (in different versions), an ejaculation from which the god Kartikeya later springs.
55. Between a third and a half of the *Manusmriti* appears also in the *Mahabharata*.
56. See my essay, "Vatsyayana's *Kamasutra*," *SSLI*, 46–53.
57. Vaman Shivram Apte, *The Student's Sanskrit-English Dictionary* (Delhi: Motilal Banarsidass, 2000), 319.

58. Alain Danielou, *The Complete Kama Sutra* (Rochester, Vermont: Park Street Press, 1994); Devdutta Sastri, *Kamasutram* (Varanasi: Chowkhamba Sanskrit Series, 1964); Parasnath Dwivedi, *Kamasutram* (Varanasi: Chowkhamba Surbharati, 1999); Wendy Doniger and Sudhir Kakar, *Kamasutra* (Oxford: Oxford University Press, 2002).
59. See Vanita, "The *Kamasutra* in the Twentieth Century," in Vanita and Kidwai, *SSLI*, 236–240.
60. Sastri, *Kamasutram,* 370–371.
61. Mary John and Janaki Nair, "Introduction," *A Question of Silence?: The Sexual Economies of Modern India* (New Delhi: Kali for Women, 1998), 1–51.
62. J. Rajasekharan Nair, "An Unusual Love Story," *Savvy*, February 2004, p. 15.
63. Peter Brown, *The Body and Society: Men, Women and Sexual Renunciation in Early Christianity* (New York: Columbia University Press, 1998).
64. See, among many others, Edward Carpenter, *Iolaus: An Anthology of Friendship* (London: George Allen & Unwin, 1902); John Boswell, *Christianity, Social Tolerance, and Homosexuality* (Chicago: Chicago University Press, 1980); Lillian Faderman, *Surpassing the Love of Men: Romantic Friendship and Love between Women* (London: Women's Press, 1981); Terry Castle, *The Literature of Lesbian*ism (New York: Columbia University Press, 2003); Joseph Pequigney, *Such is My Love* (Chicago: Chicago University Press, 1985).
65. " 'Girlfriends' in Suicide Pact,"*Blitz*, July 11, 1980, p. 7.
66. Dinesh, "Do Ladkiyon ki Dilchasp Shadi," *Madhur Kathayen*, translated from Hindi.
67. Irshad Manji, *The Trouble with Islam* (New York: St. Martin's, 2004).
68. "Bhopal Woman wants to live with her Girlfriend," *Hindustan Times*, August 19, 2004.
69. Ashis Nandy argues that village and city meet in the slum. Introduction, *The Secret Politics of Our Desires*, ed. Nandy (New Delhi: Oxford University Press, 1998), 1–19.
70. "*Sara zamana haseenon ka deewana.*"
71. Alfred Kinsey, Wardell Pomeroy and Clyde Martin, *Sexual Behavior in the Human Male* (Philadelphia: W.D. Saunders, 1948).
72. "Wo'man' Abducts Her Lady Love," Press Trust of India report in *The Times of India*, times of India.indiatimes.com/articleshow/596686.cms, April 2, 2004.
73. In the story of Supriya recounted in chapter 7, Supriya's sons addressed her co-wife (who became her lover) in Marathi as *Kaki* or Aunt.
74. Terry Castle, *Kindred Spirits* (New York: Columbia University Press, 1996); Martha Vicinus, *Intimate Friends: Women who Loved Women 1778–1928* (Chicago: University of Chicago Press, 2004).

CHAPTER 2 WHO DECIDES?: MARRIAGE LAW, THE STATE, AND MUTUAL CONSENT

1. Andrew Sullivan, *Same-Sex Marriage, Pro and Con* (New York: Random House, 1997), 67.

2. "She Changed her Sex to Marry her," *Times of India*, March 3, 1989. Manju became Manish.
3. See "Cops tell lesbian couple not to marry," *The Indian Express*, April 17, 1993, and "The Bold, Beautiful and the Damned," *The Indian Express*, April 18, 1993.
4. Suchandana Gupta, "Husband at Home, in Sari at Work," *The Telegraph* (Calcutta) May 29, 2001.
5. Divorce was impossible in England till 1857 and also impossible in colonial America. See Lawrence Friedman, *A History of American Law* (New York: Simon & Schuster, 1973), 202–212.
6. T.A. Lacey, *Marriage in Church and State* (London: Robert Scott, 1912), 44.
7. S.K. Mitra, *Hindu Law* (Delhi: Orient Publishing Co., 2000), 516–517.
8. The Sanskrit word *gandharva* refers to heavenly beings and their supposed practices.
9. See Sudarshan Dhar and M.K. Dhar, *Evolution of Hindu Family Law* (Delhi: Deputy Publications, 1986), chapters 2 and 3.
10. Dr. K.P.A. Menon, ed. *Complete Plays of Bhasa* (Delhi: Nag Publishers, 1996), Vol. III, 314.
11. Anu and Giti, "Inverting Tradition: The Marriage of Lila and Urmila," in *A Lotus of Another Color: An Unfolding of the South Asian Gay and Lesbian Experience* ed. Rakesh Ratti (Boston: Alyson Publications, 1993), 81–84.
12. Priya Solomon, "Women in Love," *The Week*, 16–18.
13. For the legal history in Canada, see Kevin Alderson and Kathleen Lahey, *Same-Sex Marriage: The Personal and the Political* (Toronto: Insomniac Press, 2004).
14. Lawrence Stone, *Uncertain Unions* (Oxford: Oxford University Press, 1992), 17.
15. Sullivan, *Same-Sex Marriage*, 105.
16. Mitra, *Hindu Law*, 645.
17. Monmayee Basu, *Hindu Women and Marriage Law* (Delhi: Oxford University Press, 2001), 40–44.
18. Mitra, *Hindu Law* 1–7.
19. Ibid., 16–17.
20. Ibid., 1.
21. While feminists want to free Muslim women from supposedly oppressive practices such as polygamy, some Hindus see the retention of polygamy in Muslim law as evidence of the government's unfair pandering to the Muslim vote-bank. One solution is internal reform by Muslims, which is beginning to happen. See Ziauddin Sardar, "Can Islam Change?" *New Statesman*, September 13, 2004.
22. Basu, *Hindu Women and Marriage Law*, 44.
23. J. Duncan and M. Derrett, *The Death of a Marriage Law* (New Delhi: Vikas, 1978), 32.
24. Mitra, *Hindu Law*, 644.
25. Dharmendra Rataul, "Even Death Can't do Us Apart," *Indian Express*, December 11, 2004.
26. Nirjhar Dixit, "Bride Grooms Bride," *Savvy*, February 21, 1988.
27. "Gay Youths Thwarted after Entering into Wedlock," March 1, 2004. www.ushome.rediff.com/news/2004/mar/01up.htm.
28. Javed Rana, "Gays' Love and Maulvi Disco," *International Khyber Mail*, October 2, 1999.

29. The Committee on the Status of Women in India, 1975, found that 5.7% of Indian Muslim men and 5.8% of Indian Hindu men had bigamous marriages.
30. Bhaurao versus State of Maharashtra, All India Records, 1965; Supreme Court Records, page 1564, Justice J. Raghubar Dayal.
31. Dhar and Dhar, *Evolution of Hindu Family Law*, 162–176.
32. See Mark Strasser, *Legally Wed* (Ithaca: Cornell University Press, 1997), 70–71 for an elaboration of this argument.
33. www.gaylawnet.com/news/1998/pa980709.htm#lesbians_wed.
34. Headline, *Times of India*, February 23, 1988, p. 4.
35. "Bride Grooms Bride," *Savvy*, February 21, 1988.
36. N.B. Grant, "Same Sex Marriage," *The Statesman* (Calcutta), March 13, 2004.
37. *Indian Express* (Bangalore) May 7, 1988.
38. "Two Girls Tie the Nuptial Knot," *Times of India*, April 20, 1998.
39. Gupta, "Husband at Home" in *The Telegraph*.
40. Ibid.
41. 1 (1981) D.M.C. 138 (Delhi), quoted in Justice S.C. Jain, *The Law Relating to Marriage and Divorce* (1979; Delhi: Universal Law Publishing Co., 1996), 73.
42. Uday Mahupkar, "Gender Jam: Case of a Curious Marriage," *India Today*, April 15, 1990, p. 10.
43. Ibid.
44. http://slate.msn.com/id/2096495/ Brendan I. Koerner, "Same-Sex Marriage for Transsexuals?" posted Tuesday, March 2, 2004, accessed March 11, 2004.
45. *The Advocate* October 10, 2000, www.advocate.com.
46. "Court Rejects Transsexual Marriage," United Press International, March 16, 2002. www.newsmax.com/archives/articles/2002/3/15/152012.shtml.
47. Anne Lamoy, "Court Overturns Ruling that Voided Transsexual's Marriage," www.ntac.org/news/01/05/18ks.html.
48. Ash Kotak, *Hijra* (London: Oberon, 2000).
49. Prannoy Bramhachari, "30-year old Oriya Eunuch Weds Boy," *Asian Age*, September 29, 2002.
50. "She Changed Her Sex to Marry Her," *Times of India*, March 3, 1989.

Chapter 3 Is the Spirit Gendered?: Fluid Gender, Sex Change, and Same-Sex Marriage

1. Chinu Panchal, "Wedded Women Cops to Challenge Sack," *Times of India*, February 23, 1988.
2. Plato, *Symposium and Phaedrus* translated Benjamin Jowett (New York: Dover, 1993), 16.
3. *The Kurma Purana*, translated Ganesh Vasudeo Tagare (Delhi: Motilal Banarsidass, 1981), Part I, 84.
4. Ibid., 92.
5. Ibid., 111.
6. Ibid., 91.

7. *The Brahmanda Purana* translated Ganesh Vasudeo Tagare (Delhi: Motilal Banarsidass, 1984), Part IV, 1034.
8. Tagare, *Kurma Purana* Vol. 2, 348.
9. Henry Vaughan, "The World," *The Norton Anthology of Poetry* (New York: W.W. Norton, 1975), 381.
10. Plato, *Symposium and Phaedrus,* 17–18.
11. Many texts emphasize this oneness. See, for example, Tagare, *Kurma Purana* Vol. 2, 385: "There is no doubt in this that he who is Isvara [Shiva] is Narayana [Vishnu]" (II.11.111).
12. There are several versions of this story. For Kumkum Roy's translation of the *Bhagavata Purana* version, see Vanita and Ruth *SSLI*, 69–71. For another version, see Tagare, *The Brahmanda Purana* (chapter X), Vol. 4, 1065–1073.
13. Serena Nanda, *Neither Man Nor Woman* (Belmont, CA: Wadsworth, 1990), gives more credence to the link than the evidence warrants. Walter Penrose ("Hidden in History: Female Homoeroticism and Women of a 'Third Nature' in the South Asian Past," *Journal of the History of Sexuality* 10:1 (2001), 3–39), follows her and also borrows many inaccuracies from Giti Thadani (*Sakhiyani: Lesbian Desire in Ancient and Modern India* [London: Cassell, 1996]), in the process confusing Hindu and Islamic provenances and terminologies, referring, for example, to the women's quarters in ancient Indian texts as *zenana.*
14. See Alf Hiltebeitel, *The Cult of Draupadi Mythologies: From Gingee to Kuruksetra* (Chicago: University of Chicago Press, 1988), especially Chapter 15, "Aravan's Sacrifice."
15. Eva Bell, "The Sisterhood of Man," *Deccan Herald*, April 19, 1998.
16. For their ritual significance, especially in Tamil Nadu, see Saskia C. Kersenboom-Story, *Nityasumangali: Devadasi Tradition in South India* (Delhi: Motilal Banarsidass, 1987).
17. Jogan Shankar, *Devadasi Cult: A Sociological Analysis* (New Delhi: Ashish Publishing, 1990), 101–112.
18. Ibid., 17–18.
19. S. Seethalakshmi, "Devadasis substitute one evil for another," *Times of India* (Bangalore), January 25, 1998.
20. Ibid.
21. K.C. Tarachand, *Devadasi Custom* (New Delhi: Reliance, 1991), 87–88.
22. *Santi Parva* I; IX, 326.
23. Diana L. Eck, *Encountering God: A Spiritual Journey from Bozeman to Banaras* (Boston: Beacon Press, 1993; 2003), 136.
24. *Bhagavata Purana*, translated G.V. Tagare (Delhi: Motilal Banarsidass, 1976), Vol. 8, Part II, p. 607.
25. See my essay, "The Self is Not Gendered: Sulabha's Debate with King Janaka," *NWSA Journal* 15:2 (Summer 2003), 76–93.
26. See Ovid, *Metamorphoses* (Philadelphia: Paul Day Books, 1999), 9: 666–797.
27. For a translation, see Vanita and Kidwai, *SSLI*, 72–76.
28. For a translation, see Vanita and Kidwai, *SSLI*, 31–36.
29. Devika Rani, "Manhood Problems," *The Week*, July 2, 1989, 47–48.

30. Chapters 87–90 in "Uttara Kanda" in *Valmiki Ramayana*. In the *Matsya Purana* version of this story, Shiva, to please Parvati, wills that everything in the forest should turn female. This text does not explicitly say that Shiva turns female.
31. For an analysis of this friendship, see Vanita and Kidwai, *SSLI*, 3–9.
32. As Julian F. Woods notes, Parikshit is "the perfect monarch embodying the qualities of both Arjuna and Krsna," *Destiny and Human Initiative in the Mahabharata* (Albany: SUNY Press, 2001), 12.
33. Ganguly, *Mahabharata* XII, 134.
34. See, for example, Richard Rambuss, *Closet Devotions* (Durham, N.C.: Duke University Press, 1998).
35. See R.C. Zaehner, *Hindu and Muslim Mysticism* (1994; Delhi: Research Press, 1999), especially the first three chapters.
36. Deshpande, *Padma Purana*, Vol. VI, p. 1962.
37. For a translation of an extract from this story, see Vanita and Kidwai *SSLI*, 90–93, 93.
38. Deshpande, *Padma Purana*, Vol. VI, 1967.
39. Reprinted as "Off the Beaten Track," in *Dilemma and other Stories: Short Fiction by Vijay Dan Detha*, translated Ruth Vanita (New Delhi: Manushi Prakashan, 1997), 3–35. For a condensed version, see Vanita and Kidwair *SSLI*, 318–324.
40. Vanita and Kidwai, *SSLI*, 321.
41. Ibid.
42. Ibid., 323.
43. Terry Castle, *The Apparitional Lesbian* (New York: Columbia University Press, 1993) sees the lesbian as a metaphoric ghost haunting literary tradition; I suggest that tradition may be the friendly ghost supporting lesbian existence.
44. Shohini Ghosh ("Queering the Family Pitch," *The Little Magazine* November–December 2000, 33–45) sees the ghost as "rearticulated man" holding out a vision of a "utopian future." I agree, but view that future as drawing from the past.
45. Vanita and Kidwai, *SSLI*, 324.

CHAPTER 4 "IMMORTAL LONGINGS": LOVE-DEATH, REBIRTH, AND UNION THROUGH LIFE AFTER LIFE

1. In two articles, "Same-Sex (female) Lovers Commit Suicide," and "Lesbian Suicides Continue," in *Sameeksha*, a Malayalam fortnightly (June 28–July 11, 1998 and January 1–15, 2000), K.C. Sebastian lists couples who committed suicide in the state of Kerala. Sahayatrika, a lesbian organization in Kerala, is now preparing an updated report. See Devaki Menon, "*Sahayatrika: Chila Anubhavapaadhangal*," in *Mythyakalkkappuram: Swavargga Laingikatha Keralathil* ed. Reshma Bharadwaj (Kottayam: DC Books, 2004), 20–35. The publicity received by these cases has overshadowed the fact that such suicides occur throughout India.
2. For one account of the emotional damage inflicted by family on an economically independent 26-year-old, see Stuart Howell Miller, *Prayer Warriors: The True*

Story of a Gay Son, his Fundamentalist Christian Family, and their Battle for His Soul (Los Angeles: Alyson Books, 1999).
3. J. Rajasekharan Nair, "An Unusual Love Story," and "I Believe," *Savvy*, February 2004, 12–22.
4. "Lesbian suicides" *Sameeksha*, January 1–15, 2000.
5. Jupinderjit Singh, "Police Summons Members of Industrialist's Family," *The Tribune*, December 13, 2002, www.tribuneindia.com.
6. "And in their Death they were United," *Indian Express*, April 16, 1992.
7. *For People Like Us* (New Delhi: AIDS Bhedbhav Virodhi Andolan, 1999), 21.
8. "Suicide by Young Lovers," *Times of India* (New Delhi) October 4, 2002.
9. "Parental Opposition leads to Lovers' Death," *Times of India*, February 6, 2004.
10. "Lovers Attempt Suicide," *The Telegraph* (Durgapur) December 4, 2003.
11. "Jilted Lovers Leap to Death from High-Rise," *The Hindu*, March 15, 2004.
12. D.R. Mace and V. Mace, *Marriage East and West* (Garden City, New York: Doubleday, 1960), 103.
13. "Gay Couple Stab Each Other," *News Today*, May 27, 1993.
14. *Mathrubhumi*, July 8, 1999.
15. *For People Like Us*, 42–45.
16. Ibid., 26–27.
17. "Macabre Suicide,"*India Today*, October 15, 1988.
18. Victor Lenous, " 'Girlfriends in Suicide Pact,' " *Blitz*, July 11, 1980.
19. Dharampal, *Civil Disobedience and Indian Tradition* (Varanasi: Sarva Sewa Sangh Prakashan, 1971).
20. Thich Nhat Hanh, *Love in Action: Writings on Nonviolent Social Change* (Berkeley, CA: Parallax Press, 1993), 43.
21. "Nurse's Love for Lady Doc Drives Sister to Suicide," *Indian Express*, September 21, 2002. www.indianexpress.com/archive_frame.php.
22. *Lokmat*, February 4, 2004.
23. "Macabre Suicide," *India Today*, October 14, 1988.
24. Dikshya Thakuri, "Lesbianism, Still Taboo and Scary," *The Himalayan Times*, December 7, 2003.
25. *Dina Thanthi* (Bangalore edition), October 4, 2002, p. 10. Translated from Tamil by Sangama, a Bangalore based LGBT group.
26. "Lesbian Suicides,"*Sameeksha* June 1–15, 1999.
27. "Two Girls Marry Each Other, Held," *Indian Express*, April 1, 2000.
28. From Kiran's letter, printed under the title, "Abducted by her Family: Khurshid Jahan's Struggle," in *Manushi* No. 18, October–November 1983, 22–23.
29. Amy Waldman, "Broken Taboos Doom Lovers in an Indian Village," *New York Times*, March 28, 2003. See also Avijit Ghosh's article on honor killings in north India, "Whose Honor is it Anyway?" *The Telegraph*, January 18, 2004.
30. "Lovers Commit Suicide with Parent's 'No,' " *Pakistan Times*, December 17, 2003. www.pakistantimes.net.
31. " 'Lesbian' Romance on the Rocks?"*Indian Express*, September 3, 1993.
32. Waris Shah, *The Love of Hir and Ranjha*, translated Sant Singh Sekhon (Ludhiana: Old Boys' Association, College of Agriculture, 1978), 262–263.
33. Ibid., 262.
34. *For People Like Us* (New Delhi: ABVA, 1999), 21.

35. V. Radhika, "Why Can't Women Love Women?" www.indianest.com/wfs/wfs070.htm Contains interviews with members of lesbian group Olava, Pune.
36. *Dina Thanthi.*
37. From *Prem Pujari*, 1970.
38. *Kathasaritsagara, The Ocean of Story* translated C.H. Tawney (Delhi: Motilal Banarsidass, 2nd edn. 1924), 120. All further references are to page numbers in this edition, unless otherwise indicated.
39. "Girlfriends," *Blitz.*
40. Shakuntala Devi, *The World of Homosexuals* (Delhi: Vikas, 1977).
41. An earlier version of this story is found in Bhasa's play, *Avimaraka.* There, the lover attempts suicide both by fire and by jumping from a cliff. The second time, he is saved by some Gandharvas, and is restored to his adoptive father, a king. The princess also attempts suicide by hanging but is stopped in time to marry him.
42. Donald H. Shively, *The Love Suicide at Amijima* (Cambridge, MA: Harvard University Press, 1953), 25.
43. D.R. Mace and V. Mace, *Marriage East and West*, 123.
44. Nandy, *Alternative Sciences*, 116.

Chapter 5 A Second Self: Traditions of Romantic Friendship

1. Reported on the Gay Bombay list, gaybombay@yahoogroups.com, June 13, 2001.
2. See Raj Ayyar, "Yaari," in *Lotus of Another Color*, 167–174.
3. "Laelius: On Friendship," in *Cicero: On the Good Life* translated Michael Grant (London: Penguin, 1971), 191.
4. See, for example, "Jayalalithaa in Guruvayur to offer calf elephant," http://www.rediff.com, July 2, 2001.
5. Thanks to T. Muraleedharan for pointing this out, and also for sending me translations of some Malayalam news reports.
6. "50 Devotees Dead in TN Stampede," *Hindustan Times* (Kumbakonam, UNI, PTI) February 19, 1992.
7. Rediff online news magazine, http://www.rediff.com, May 3, 1998.
8. http://forumhub.com/tnhistory/4895.11.08.34.html October 10, 2002.
9. In 2004 her party lost national-level elections but is still in power in Tamil Nadu.
10. Reported by BBC, February 5, 1996.
11. Interview by M.D. Nalapat for Editor's Choice program (telecast on Doordarshan channel 2 on January 10, 1999), www.rediff.com.
12. Gayle Rubin, "The Traffic in Women," in *Toward an Anthropology of Women* ed. R. Reiter (New York: Monthly Review Press, 1976).
13. *Aristotle: The Nicomachean Ethics* translated David Ross (Oxford: Oxford University Press, 1925), 198.

14. Kidwai, "Introduction: Medieval Materials in the Perso-Urdu Tradition," *SSLI*, 107–125.
15. Plato, *Symposium and Phaedrus*, 32.
16. Aristotle, *Nicomachean Ethics*, 200.
17. Ibid., 244.
18. "Laelius" in *Cicero*, 184, 227.
19. *The Confessions of St. Augustine*, translated Rex Warner, introduced Vernon J. Bourke (Harmondsworth: Penguin, 1963), 74.
20. Visnu Sarma, *The Pancatantra*, translated Chandra Rajan (New Delhi: Penguin, 1993), Introduction, xlv. All further quotations in English are from this edition. Sanskrit quotations are from the critical edition, Sudhakar Malviya, ed. *Panchatantram* (Varanasi: Krishnadas Academy, 1993).
21. *Jaina Vivaha Paddhati* (Bikaner: Premraj Bohra, 1978), 2.
22. See my essay, "Ayyappa and Vavar: Celibate Friends," *SSLI*, 94–99.
23. "Laelius," in *Cicero*, 189.
24. "Of Friendship," *Complete Essays*, 141.
25. Ibid.
26. See Scott Kugle, "Sultan Mahmud's Makeover," *Queering India*, ed. Vanita, 30–46.
27. *Bihari: The Satasai*, page 254, verse 558. See also page 92, verse 132.
28. "Of Friendship," *Complete Essays*, 139.
29. Ibid.
30. Pandit Durgaprasad and Kasinath Pandurang Parab, ed. *The Kathasaritsagara of Somadevabhatta* (Bombay: Nirnaya Sagara Press, 1852; 4th edn. 1930), 86. My translation.
31. Plato, *Symposium and Phaedrus*, 40.
32. "Of Friendship," *Complete Essays*, 138.
33. Ibid.
34. Rajan, *Pancatantra*, 209.
35. Ibid.
36. Warner, *Confessions*, 76.
37. Ibid.
38. "Of Friendship," *Complete Essays*, 143.
39. Walter Pater, *The Renaissance* (1873; London: Macmillan, 1900), 28.
40. *Madhumalati* translated Aditya Behl and Simon Weightman (Oxford: Oxford University Press, 2000), 104–105.
41. Robin Darling Young, "Gay Marriage: Reimagining Church History," *First Things* 47 (1994): 43–48.
42. See, for example, *Mahabharata* Vol. VII, p. 105, *Karna Parva* Section XLII; Vol. VIII, p. 295, *Shanti Parva* Part I, Apadharmanusasana Parva, Section CXXXVIII.
43. Nirjhar Dixit, "Bride Grooms Bride," *Savvy*, February 21, 1988.
44. Manjeet Singh, "Urmila Weds Leela, Both Get Sacked," *Sunday Observer*, February 21, 1988.
45. Chinu Panchal, "Wedded Women Cops to Challenge Sack," *Times of India*, February 23, 1988.

46. For translations, see Vanita and Kidwai *SSLI*, 233–235; 327–333; 352–355.
47. www.geocities.com/rejionnet/Home. Homepage of Reji G. Accessed December 26, 2003.
48. Chinu Panchal, "Wedded Women Cops to Challenge Sack," *Times of India*, February 23, 1988.
49. Justin Huggler, " 'Married' lesbian couple defy their families and vow to fight Indian law," *The Independent*, December 13, 2004.
50. http://edition.cnn.com/2003/WORLD/asiapcf/south/02/11/india.valentines.ap/, February 11, 2003, Bombay, "India Set for Anti-Valentine Demos."
51. People's Daily Online, http://english.peopledaily.com.cn/, February 13, 2001, "Call for Valentine's Ban in India," accessed December 14, 2003.
52. http://www.usatoday.com/news/world/2003-02-14-india-valentine_x.htm, February 14, 2003, Bombay, accessed January 5, 2004.
53. For a translation of excerpts with commentary, see Scott Kugle, "*Haqiqat al-Fuqara*: Poetic Biography of 'Madho Lal' Hussayn," *SSLI*, 145–156.
54. "The Kingdom of Silence," Lawrence Wright, *The New Yorker*, January 5, 2004, 48–73.
55. Ibid.

Chapter 6 Monstrous to Miraculous—Same-Sex Reproduction and Parenting

1. Modified by Dan Savage from Jesse Green, *The Velveteen Father: An Unexpected Journey to Parenthood* (New York: Ballantine, 2000).
2. See Seema Kumar, "God's Own," *Hindustan Times*, March 16, 1997.
3. Dayanita Singh, *Myself Mona Ahmed* (Berlin: Scalo Zurich, 2001), 83–85.
4. *Trikone*, special issue on immigration, ed. Ruth Vanita, 2004.
5. The ancient Greek Gods had no children in the regular way though their sexual pleasures were unlimited; being immortal, they did not need to replace themselves. Hindu Gods and Goddesses too do not produce children in the regular way.
6. Kaviraj Kunjalal Bhishagratna, *An English Translation of the Sushruta Samhita*, (Varanasi: Chowkhamba Sanskrit Series Vol. XXX, 1991), 132.
7. *Caraka-Samhita*, translated and ed., Priyavrat Sharma (Varanasi: Chaukhamba Orientalia, 2nd edn. 1994), Vol. 1, pp. 418–427.
8. The story of Bhagiratha's birth to two mothers also appears in later texts produced in Bengal, such as Bhavananda's *Harivansha*, Mukundarama Chakravartin's *Kavikankanachandi*, and Adbhutacharya's sixteenth century *Ramayana*.
9. See, for example, Deshpande, *Padma Purana* VI. 21. 2–8. See also Tagare, *Kurma Purana* I. 21.8–9a.
10. Tony K. Stewart and Edward C. Dimock, "Krttibasa's Apophatic Critique of Rama's Kingship," in Paula Richman, *Questioning Ramayanas: A South Asian Tradition* (Berkeley: University of California Press, 2001), 243–264.

11. For dating controversies, see *Classical Hindu Mythology: A Reader in the Sanskrit Puranas* translated and edited Cornelia Dimmitt and J.A.B. van Buitenen (Philadelphia: Temple University Press, 1978), Introduction.
12. Asoke Chatterjee Sastri, ed. *The Swarga Khanda of the Padma Purana* (Varanasi: All-India Kashiraj Trust, 1972), 43–46. My translation. The editor, in his introduction, gives an account of various recensions, and discusses the unique Bengali script version of the Bhagiratha story.
13. Shastri, *Swarga Khanda* 16:12, 141. My translation.
14. Ibid., 16: 15, 141.
15. For the dating problem of the *Krittivasa Ramayana*, see Sukumar Sen, *History of Bengali Literature* (New Delhi: Sahitya Akademi, 1960; 1979), 67–69.
16. Nandkumar Avasthi ed. *Krittivasa Ramayana*, with Hindi translation (Lucknow: Bhuvan Vani, 1966), 60–65. The story occurs in the first section, the *Adi Kanda*, which traces the ancestry of Rama.
17. Vanita and Kidwai, *SSLI*, 100–102.
18. Ibid., 101. Translation by Kumkum Roy. Thanks to Arindam Chakrabarti for explicating the connotations of some words.
19. Awasthi, *Krittivasa Ramayana* 64.
20. Nalinikanta Bhattasali, ed., *Ramayana-Adikanda* (Dacca: P.C. Lahiri, Secretary, Oriental Texts Publication Committee, University of Dacca, 1936). Here it is translated for the first time into English by Anannya Dasgupta. See Bhattasali's introduction for an account of the different manuscripts.
21. Ibid., 90–92.
22. Vanita and Kidwai, *SSLI*, 101.
23. John Boswell, *The Kindness of Strangers: the Abandonment of Children in Western Europe from Late Antiquity to the Renaissance* (New York: Pantheon Books, 1988).
24. Deshpande, *Padma Purana*, V. 15. 42b–46.
25. See "*Shiva Purana*: The Birth of Ganesha," in *SSLI*, 81–84.
26. For essays debating the question, see *Is the Goddess a Feminist?: the Politics of South Asian Goddesses* ed. Alf Hiltebeitel and Kathleen M. Erndl (New York: New York University Press, 2000).
27. For a detailed analysis, see my essay, "Sita Smiles: Wife as Goddess in the *Adbhut Ramayana*," in Ruth Vanita, *Gandhi's Tiger and Sita's Smile: Essays in Gender, Sexuality and Culture* (New Delhi: Yoda Press, forthcoming 2005).
28. Mark Jordan, *The Invention of Sodomy in Christian Theology* (Chicago: Chicago University Press, 1997) 175.
29. See Vanita, "Ayyappa and Vavar: Celibate Friends."
30. For an analysis of Vedic patterns, see Vanita and Kidwai *SSLI*, 14–17; also Giti Thadani, *Sakhiyani: Lesbian Desire in Ancient and Modern India* (New York: Cassell, 1996).
31. For an interpretation of this friction as an all-female creative force, see Lawrence Durdin-Robertson, *The Goddesses of India, Tibet, China and Japan* (Clonegal, Eire: Cesara Publications, 1976) s.v. arani.
32. J. Rajasekharan Nair, "I Believe," *Savvy* 22.
33. See *Trikone*, special issue on, "Family Values," 18:3 (September 2003).

CHAPTER 7 ALL IN THE FAMILY: SAME-SEX RELATIONSHIPS IN TRADITIONAL FAMILIES

1. P.I. Rajeev, "To save wife, Man Marries Her 'Lover,' " *Indian Express*, December 11, 2004.
2. See my essay, "The New Homophobia: Ugra's *Chocolate*," *SSLI*, 246–252.
3. Opinion issued November 7, 2002, in the matter of David Blanchflower and Sian E. Blanchflower.
4. See David Leddick, *The Secret Lives of Married Men* (Los Angeles: Alyson 2003) for interviews with 39 such men.
5. J.L. King and Karen Hunter, *On the Down Low: A Journey into the Lives of "Straight" Black Men who Sleep with Men* (Broadway Books, 2004).
6. "Sex by Whatever Name," *Psychological Foundations: The Journal* 5(1) June 2003.
7. Lillian Faderman, *Naked in the Promised Land* (Boston: Houghton Mifflin, 2003).
8. Personals in *Trikone* Magazine for South Asians, April 2002, 22.
9. Ibid.
10. Personals in *Trikone*, October 2000, 22.
11. Personals in *Trikone*, July 2001, 22.
12. From AMALG website on October 16, 2003. Page http://members.tripod.com/~Dating_Service/people.html.
13. Personals in *Trikone*, October 2002, 22.
14. Personals in *Trikone*, July 2001, 22.
15. Personals in *Trikone*, April 2001, 22.
16. Sachin Kedar, on Khush chatlist, March 30, 2003, subject "Gaybombay Thread on Marriages of Convenience."
17. Part of the advertisement cited in *Trikone*, July 2001, 22.
18. Personals in *Trikone*, October 1998, 22.
19. Personals in *Trikone*, October 1998, 22.
20. Rashmi, March 26, 2003, on khush chatlist, subject "GayBombay Thread on Marriages of Convenience."
21. Kannanodi, October 27, 2002, on khush chatlist, subject "Marriages of convenience, any thoughts?"
22. From AMALG website on October 16, 2003. Page http://members.tripod.com/~Dating_Service/people.html.
23. Personals in *Trikone*, January 2000, 22.
24. Personals in *Trikone*, October 2000, 22.
25. From AMALG website on October 16, 2003. Page http://members.tripod.com/~Dating_Service/people.html.
26. Personals in *Trikone*, July 2000, 22.
27. Michelangelo Signorile, "Point and Lick," *New York Press*, November 11, 2003.
28. Personals in *Trikone*, June 2003, 22.
29. Ibid.
30. Personals in *Trikone*, October 2000, 22.
31. Personals in *Trikone*, October 2002, 22.

32. Supriya, as told to and translated from Marathi by U.S.G., in *Facing the Mirror*, 121–124. The names are possibly pseudonyms.
33. The formation is the same in Sanskrit—*sapatni* is a co-wife and the relationship between co-wives' children is *sapatnam*; this word acquired the secondary meaning of "rival."
34. K.P.A. Menon, ed. *Complete Plays of Bhasa* (Delhi: Nag Publishers, 1996), 199.
35. English text *The Katha Sarit Sagara* translated C.H. Tawney (1880; New Delhi: Munshiram Manoharlal, 1968). All page numbers refer to this edition. Sanskrit text, *The Kathasaritsagara of Somadevabhatta*, ed. Pandit Durgaprasad and Kasinath Pandurang Parab (Bombay: Nirnaya Sagara Press, 4th edn., 1930), I. XVI.18.
36. See my 1990 essay, " 'Shall We Part, Sweet Girl?' The Role of Celia in *As You Like It*," reprinted in *Critical Theory, Textual Application* ed. Shormishtha Panja (New Delhi: Worldview Press, 2002).
37. Meera, "Finding Community," *A Lotus of Another Color* ed. Rakesh Ratti (Boston: Alyson Publications, 1993), 234–245.
38. "Lesbian Couple Split by Police," *Indian Express*, June 7, 1997.
39. *For Straights Only* (2001), documentary film by Vismita Gupta-Smith.
40. See interview with Lakshmi and Arka, Indians living in the United States, both bisexual and non-monogamous, and married to each other by choice, "Extended Family," in *Lotus of Another Color*, 265–278.
41. *Trikone*, September 2003, "Family Values" issue, carries interviews with a man and woman who have two children from a 22-year MOC, but now are divorced and living with same-sex partners.
42. Kamala Das, *The Sandal Trees* (Delhi: Orient Longman, 1995).
43. For a translation, see Vanita and Kidwai, *SSLI*, 304–310.
44. Anita Nair, *Ladies Coupe* (New Delhi: Penguin, 2001), 219.

CHAPTER 8 "MARRIED AMONG THEIR COMPANIONS": FEMALE-FEMALE RELATIONS IN PREMODERN EROTICA

1. Terry Castle, *The Literature of Lesbianism* (New York: Columbia University Press, 2003) is among the few texts to counter this bias, including writings about lesbianism by both men and women.
2. Kidwai and I translated and analyzed some of these poems in *SSLI*, 191–194; 220–228.
3. See Mohammed Sadiq, *A History of Urdu Literature* (1964; Delhi: Oxford University Press, 1995).
4. For critical commentary in Urdu, see Sayyid Sibt-i Muhammad Naqvi, *Intikhab-i Rekhti* (Lucknow: Uttar Pradesh Urdu Akademi, 1983), Tamkin Kazmi, *Tazkirah-i Rekhti* (Hyderabad: n.p., 1930), Khalil Ahmed Siddiqui, *Rekhti ka Tanqidi Mutalia* (Lucknow: Nasim Book Depot, 1974).
5. Two verses by Naubahar, pen-name "Zaleel" are included in Sayyid Sibt-i Muhammad Naqvi's anthology, *Intikhab-i Rekhti* (Lucknow: Uttar Pradesh

Urdu Akademi, 1983), 74. He describes her as a *kaneez* (maidservant) of Prince Sulaiman Shikoh. C.M. Naim, "Transvestic Words?: The Rekhti in Urdu," *The Annual of Urdu Studies* No. 16 (2001), 3–26, mentions two other female poets (footnote 3).

6. For this campaign in Urdu literature, see Frances Pritchett, *Nets of Awareness: Urdu Poetry and its Critics* (Karachi: Oxford University Press, 1995). For its effects on homoeroticism in Indian literature, see "Introduction: Modern Indian Materials," *SSLI*, 236–240; and Scott Kugle, "Sultan Mahmud's Makeover," *Queering India*, 30–46.
7. John T. Platts, *A Dictionary of Urdu, Classical Hindi and English* (1884; Delhi: Munshiram Manoharlal, 1997), 421.
8. Ibid., 421.
9. Veena Talwar Oldenburg, "Lifestyle as Resistance: The Case of the Courtesans of Lucknow," in *Lucknow: Memories of a City* ed. Violette Graff (Oxford: Oxford University Press, 1997), 136–154; 149.
10. Havelock Ellis, *Studies in the Psychology of Sex* (1900; New York: Random House, 1942), I, 208–09.
11. Platts, *Dictionary of Urdu and English*, 530.
12. S.S.M. Naqvi, ed. *Intikhab-e Rekhti* (Lucknow: Urdu Academy, 1983), 13, *ghazal* 15. All further citations from this anthology are indicated in parentheses thus: IR, 13:15, where the first number is the page number and the second number that of the *ghazal*. When the *ghazal* has no number, only the page number is given.
13. *Diwan-e Rangeen*, pp. 5–6, ST 9471, P964, in Raza Memorial Library, Rampur.
14. Excerpted in Sabir Ali Khan, *Sa'adat Yar Khan Rangeen* (Karachi: Anjuman-e Taraqqi-e Urdu, 1956), 415.
15. Ibid., 416.
16. Ibid., 412.
17. *Selected Satires of Lucian*, translated and ed. Lionel Casson (New York: W.W. Norton, 1962), 303–305; 304.
18. *Kalam-e Insha* (section *Diwan-e Rekhti*) ed. Mirza Mohammad Askari and Mohammad Rafi Fazal Deoband (Allahabad: Hindustani Akademi Uttar Pradesh, 1952), 416, *ghazal* 43. All citations to Insha's poems from this collection are indicated thus: KI, 406–07: 21, with the first number indicating the page and the second the *ghazal* number.
19. Vanita and Kidwai, *SSLI*, 268, 292, 299, 314.
20. From Vanita and Kidwai, *SSLI*, 227.
21. Ibid.
22. See Waris Shah, *Love of Hir and Ranjha*, 198–202. A parallel may be found in the *querelle des femmes* or "woman question" debate in medieval European poetry.
23. For the original *Chaptinamas*, see Iqtida Husain, ed. *Kulliyat-e Jur'at* (Napoli: Istituto Universitario Orientale, 1971), II: 261–262; 294–297. For a complete translation by Saleem Kidwai, versified by Ruth Vanita, see *SSLI*, 222–225.
24. Platts, *Dictionary of Urdu, Hindi and English*, 782.
25. *Diwan-e Rangeen*, ST 9471, P964, p. 31.
26. Naim, "Transvestic Words?" footnote 42.

27. *Diwan-e Rangeen*, ST 9471, P964, 294.
28. See Vanita and Kidwai, *SSLI*, 127, 143–156, 161–174 for analysis and examples.
29. See Tariq Rahman, "Boy-Love in the Urdu Ghazal," *Annual of Urdu Studies* 7 (1990): 1–20; C.M. Naim, "The Theme of Homosexual (Pederastic) Love in Pre-Modern Urdu Poetry," in *Studies in the Urdu Ghazal and Prose Fiction* Muhammad Umar Memon ed. (Madison: University of Wisconsin, 1979), 120–142; Saleem Kidwai, "Introduction: Medieval Materials in the Perso-Urdu Tradition," in *SSLI*, 107–125.
30. For these and other conventions of the Urdu *ghazal*, see Pritchett, *Nets of Awareness*, 77–122.
31. Published in London from 1879 to 1880. Reprinted as *The Pearl* (New York: Blue Moon Books, 1996).
32. Danielou, *Kama Sutra*, 192.
33. Oldenburg, "Lifestyle as Resistance,"149.
34. See Kidwai, "Medieval Materials in the Perso-Urdu Tradition," *SSLI*, 107–184.
35. See Ashis Nandy, *Intimate Enemy* and Mrinalini Sinha, *Colonial Masculinity* (Manchester: Manchester University Press, 1995).
36. Muhammad Husain Azad, *Ab-e Hayat* translated and ed. Frances Pritchett (New Delhi: Oxford University Press, 2001), 232–233.
37. Carla Petievich, "Doganas and Zanakhis: The Invention and Subsequent Erasure of Urdu Poetry's 'Lesbian' Voice," in *Queering India* ed. Vanita, 47–60; C.M. Naim, "Transvestic Words?"
38. Friedhelm Hardy, *Viraha Bhakti* (Delhi: Oxford University Press, 1983), 307–371.
39. Petievich, "Doganas and Zanakhis,"; Naim, "Transvestic Words?"
40. Oldenburg, "Lifestyle as Resistance," 136.
41. From *Tazkirah khush ma'arika-i ziba*, quoted in Kathryn Hansen, "The Indar Sabha Phenomenon," in *Pleasure and the Nation* ed. Rachel Dwyer and Christopher Pinney (Delhi: Oxford University Press, 2001), 76–114.
42. See Frances Pritchett, "Constructing a Literary History, a Canon, and a Theory of Poetry," introductory essay in *Ab-e Hayat*, especially 38–40.
43. Quoted from poet Insha's *Darya-e Latafat* (a book on the history of Urdu) in *Ab-e Hayat*, 119–120.

Chapter 9 Aspiring to Union: Twentieth-Century Cinema

1. "Chup hai dharti," *House No. 44* (1955).
2. From the epistles of Greek sophist Philostratus (ca. AD 170–245), and the poems of Roman poet Catullus (born 82 BC).
3. Shohini Ghosh, "*Queer Pleasures for Queer People*," in *Queering India*, 208.
4. See Vinay Lal, "The Impossibility of the Outsider in the Modern Hindi Film," *The Secret Politics of Our Desires*, ed. Ashis Nandy (Delhi: Oxford University Press, 1998), 234–265.

5. Interview with Barbra Streisand, *The Advocate*, October 28, 2003, 44–45.
6. Shohini Ghosh, "The Importance of Being Madhuri," *Zee Premiere* December 2000, 164–169.
7. "Thandi hava kali ghata," *Mr and Mrs 55* (1955).
8. "Raat akeli hai," *Jewel Thief* (1967).
9. "Voh chand khila," *Anari* (1959).
10. "Ai mere zindagi," *Taxi Driver* (1954).
11. "Janey kahan mera jigar," *Mr and Mrs 55* (1955).
12. See Ghosh's analysis of a couple of recent songs of this kind, "Queer Pleasures," in *Queering India*.
13. Ibid.
14. See Aniruddh Chawda, "Rattling the Prison Cage," *Trikone* 19:2 (2004): 6–8; Ziya-us Salam, "Bold but Cliched," *The Hindu*, June 18, 2004; and, for somewhat earlier films, "A Touch of Realism," *Hindustan Times* Infotainment, November 15, 1996.

CHAPTER 10 CONCLUSION

1. Rajiv Malik, "Discussions on Dharma," *Hinduism Today*, October–November–December 2004, 30–31.
2. Prathima Nandakumar, "Gay Marriages Groom Anti-AIDS Battle," *Times News Network*, August 28, 2004, www1.timesofindia.indiatimes.com/articleshow/830976.cms. See also Lekha Menon, "I am proud, I am gay," *Times News Network*, December 10, 2001, timesofindian.indiatimes.com/articleshow/1249531931.cms.
3. Rajiv Malik, "Discussions on Dharma," *Hinduism Today*, October–November–December. 2004, 30–31.
4. Interview with Jim Gilman by Arvind Kumar, *Trikone* 11:3 (July 1996), 6–7.
5. Letter to Amara Dasa, dated July 7, 2002, www.galva108.org/perspectives3.html.Accessed August 31, 2004.
6. Condensed from a conversation with Swami Bodhananda, August 30, 2004.
7. Sonya Jones, "Gurumayi Chidvilasananda: A Revisionary Feminist Reading," paper presented at American Association of Religion, Atlanta, March 8, 2002.
8. Malik, "Discussions on Dharma," 31.
9. www.galva108.org/perspectives1.html, accessed August 31, 2004.
10. Prabhjot Singh, "Canadian MP Bains backs same-sex marriage," Tribune News Service, Chandigarh, January 24, 2005.
11. www.sikhnet.com. Questions in Youth Forum, July 15, 2002, September 9, 2003, and November 5, 2000.
12. Some Muslims based in the United States and Canada have founded the Progressive Muslims Union, which takes a pro-gay stance. See www.muslimwakeup.com; Al Fatiha is an organization of LGBT Muslims. See www.alfatiha.net/
13. *Less than Gay* (New Delhi: AIDS Bhedbhav Virodhi Andolan, 1991), 6–7.

14. Louie Crew, Reverend Kaeton and Maxine Turner, "Two Grooms," in *A Sea of Stories* ed. Sonya Jones (New York: Haworth, 2000), 183–198.
15. Cited from www.norodomsihanouk.info, in Frida Ghitnis, "Cambodia: The king speaks out in favor of gay marriage," March 14, 2004. www.miami.com/mld/miamiherald/news/opinion/8172427.htm?template=contentModules/printstory.jsp.
16. See Aditya's account of his wedding, *Trikone* 15:2 (April 2000), 12–13.
17. Telephone conversation with Arvind, September 13, 2004.
18. Arvind Kumar, "A Gem of a Life," *Trikone* 19:1 (2004), 6–7.
19. "Gay Wedding Puts Spotlight on Indian Laws," by Frederick Noronha, *Indo-Asian News Service*, January 7, 2003; see also Georgina L. Maddox "Illegal here, so Paris seals Gay Love in Goa," *Indian Express*, December 30, 2002; "Out and Proud and Married," *Deccan Herald*, December 28, 2002.
20. Aman Sharma, "Women in Love, Marry Each Other, Enter Suicide Pact," *Indian Express*, May 10, 2005. http://www.indianexpress.com/full_story.phip?content id=70060.
21. His own translation, from the wedding program.

APPENDIX: NOTE ON THEORY AND METHODOLOGY

1. Rajat Kanta Ray, *Exploring Emotional History: Gender, Mentality and Literature in the Indian Awakening* (New Delhi: Oxford University Press, 2001), 38.
2. *Cousine Bette*, chapter 37.
3. Rajat Kanta Ray, *Exploring Emotional History*, xiii.
4. Sudhir Kakar, *Intimate Relations: Exploring Indian Sexuality* (New Delhi: Penguin, 1989), especially 1, 22, 41, 82–83.
5. Castle, Introduction, *The Literature of Lesbianism* 1–56; 6.
6. See, for the West, Bernadette Brooten, *Love between Women: Early Christian Responses to Female Homoeroticism* (Chicago: Chicago University Press, 1996); John Boswell, "Categories, Experience and Sexuality," *Forms of Desire: Sexual Orientation and the Social Constructionist Controversy* ed. Edward Stein (New York: Garland Publishing, 1990); Terry Castle op. cit.; Emma Donoghue, *Passions between Women*; and for India, Kidwai's and my work in *SSLI*, and also Michael J. Sweet and Leonard Zwilling, "The First Medicalization: The Taxonomy and Etiology of Queers in Classical Indian Medicine," *Journal of the History of Sexuality* 3:4 (1993): 590–607, and " 'Like a City Ablaze': The Third Sex and the Creation of Sexuality in Jain Religious Literature," *Journal of the History of Sexuality* 6:3 (1996): 359–384.
7. Ruth Vanita, "Gandhi's Tiger: Multilingual Elites, the Battle for Minds, and English Romantic Literature in Colonial India," *Postcolonial Studies* 5:1 (2002), 95–112.
8. "It's Not Natural," *Guardian* July 3, 2004, www. guardian.co.uk/gayrights/story/0.12592.1253130.00.html.
9. "Lesbian Group in Kerala School," *Indian Express*, Madurai, January 29, 1992.

10. For such narratives by Indians, see Rakesh Ratti, ed. *A Lotus of Another Color* (Boston: Alyson Publications, 1993; Hoshang Merchant, ed. *Yaraana: Gay Writing from India* (Delhi: Penguin, 1999); Ashwini Sukhthankar, ed. *Facing the Mirror: Lesbian Writing from India* (Delhi: Penguin, 1999).
11. See Vanita and Kidwai *SSLI*, 121–124.
12. Sastri, *Kamasutram*, 301. Sastri wrongly transcribes the cerebral "t" as a dental "t"; Doniger compounds the error by translating and explaining the word as "*chapaathi*," or bread (Doniger and Kakar, *Kamasutra*, xxxv). See chapter 8 for analysis of the *rekhti* poem from which Sastri quotes this line.
13. See Ashis Nandy, *The Intimate Enemy* (Delhi: Oxford University Press, 1983), and *Rajat Kanta Ray, The Felt Community* (New Delhi: Oxford University Press, 2003), 3–172; also Vanita and Kidwai, *SSLI*, xv.
14. Ruth Vanita, "Gandhi's Tiger."
15. *Less than Gay* (New Delhi: ABVA, 1991), 93.
16. See Arvind Narrain, *Queer: Despised Sexuality, Law and Social Change* (Bangalore: Books for Social Change, 2004), for an account of struggles around the law launched by Indian LGBT and human rights groups, and for a listing of these groups.
17. Ashis Nandy, *Alternative Sciences: Creativity and Authenticity in Two Indian Scientists* (1980; New Delhi: Oxford University Press, 2003), 120–121.
18. Lawrence Stone, *The Family, Sex and Marriage in England 1500–1800* (London: Weidenfeld & Nicolson, 1977), 56.

Index

Pages where major themes are discussed in detail appear in bold print.